Physics With Robotics

An NXT and RCX Activity Guide

For Secondary and College Physics

William Church
Tony Ford
Natasha Perova

College House Enterprises, LLC
Knoxville, Tennessee

Version 1.0: May, 2009

This manuscript was prepared using Word 2008 with 12 point Times New Roman font. Publishing and Printing Inc., Knoxville, TN printed this book.

College House Enterprises, LLC.
5713 Glen Cove Drive
Knoxville, TN 37919, U.S.A.
Phone or FAX (865) 558-6111
Email jwd@collegehousebooks.com
http://www.collegehousebooks.com

Thirteen digit ISBN 978-0-9792581-4-5

Foreword

My eldest son was not sure what he wanted to do in college until he took physics in his senior year in high school. His physics class got him excited about science, got him to appreciate calculus, and gave him a major. So why did it take him 12 years of schooling to get to a class that exposed him to physics? Although there had been a few physics units along the way (simple machines, light and sound, and so on), they tended to be few and far between and unconnected. The usual argument is that physics and calculus work so naturally together that there is no reason to teach one without the other. To me, that is like arguing that we should not teach biology until students understand exponential growth and chemistry — yet biology is in every 1st grade class. Although calculus does a wonderful job of modeling physical events, I do not believe it is necessary to fundamentally understand what a force and a torque are, or what current and light are. Why is it that most high school graduates have not heard of torque (except as a movie title), yet take advantage of it every time they use a stapler, mount a bicycle, or open a door? Even in college, angular momentum is seen as far more difficult than linear momentum. Why? I think what we need is continual interaction with these phenomena and a strong sense of curiosity and self-confidence on the part of the student. Tony, Bill, and Natasha have combined over 20 robotics-based activities to awaken that curiosity and enhance that understanding for a wide range of mathematical skill levels.

As an engineer, I am particularly excited with their engineering approach. A number of students have difficulty learning math or science purely for the sake of the knowledge. Engineering problems give them a reason to learn the material and an opportunity for them to transfer that knowledge to new situations. As more and more research shows that we all learn differently (and often not in the same way that the teacher learns), we are starting to realize that we (teachers) must present material in multiple ways. Further, as teachers, we try and understand the models our students have formed in their head as to why things happen the way they do. Students need multiple ways to express these models to the teacher: writing, math, construction, or verbal debate are some possible ways. Misconceptions in these models abound and research is showing that it can be very difficult to change these misconceptions - sometimes to the point where the student actually remembers (falsely) the teacher teaching the misconception. Natasha, Tony, and Bill have pulled together a number of great activities that are meant to extend classroom work by giving the students another method for expressing these models and for sparking discussion between students on the validity of these models. They have more than 50 years of teaching experience between them and I have found them to be very innovative, creative, and successful teachers.

I think you will find that these activities would work well in my son's physics class, a middle school science class, in the elementary school, or even in the living room as parents and children investigate together. In all cases, I am convinced these activities will spark excitement and motivation in students; some whom have failed to participate in the more traditional learning environment. I believe we learn best when we are using our hands, moving around, arguing amongst each other, and generally being curious - all promoted by these activities. The true strength of the activities, though, lies in the way that we catalyze the learning - through questions, discussions, and reports. I hope that this book will inspire more teachers to bring engineering-based physics into their classroom, to help students drive their own learning, and to motivate fellow teachers to share their classroom experiences, successes and failures, with others through conferences, websites, and more books.

Chris Rogers
Professor Department of Mechanical Engineering
Director, Center for Engineering Education and Outreach
Tufts University

About the Authors

William Church

William received his Bachelor of Science in Physics from Binghamton University in 1992. He received his Master of Arts in Teaching from Cornell University in 1997. He began his teaching career in Dover, New Hampshire, USA in 1997 and currently teaches Physics, Physical Science, and Robotics in Littleton, NH. William has been using engineering design challenges and digital technology tools in his physics curriculum since 1997.

In 2003, William began consulting with the Tufts Center for Engineering Education and Outreach (CEEO). With Tufts, he works on developing activities, giving presentations, and leading teacher workshops related to using LEGO® MINDSTORMS® in secondary level physics classes. He also works on projects such as a high school to elementary school science and technology outreach model; Stop Action Movie (SAM) modeling software, hardware, and curriculum; and LABVIEW for High School. In 2006, William met fellow authors, Natasha Perova and Tony Ford. The group shared an interest and experience with utilizing LEGO MINDSTORMS in physics curricula. After conducting a presentation and workshop together at Tufts University, they began actively collaborating on this book in 2007.

William's work has been supported by funding from the Tufts CEEO, the New Hampshire Charitable Foundation's Christa McAuliffe Sabbatical program, the Lemelson-MIT program, PBS.org (WIRED Science), NH SAU 35 and 84, North Country Educational Services, Antioch New England's COSEED project, and the General Electric ELFUN Foundation.

Tony Ford

Tony has the following qualifications:

- Bachelor of Applied Science in Math and Physics at the University of Southern Queensland completed in 1974.
- Diploma of Education from University of Queensland
- Graduate Diploma in Computer Education from the Queensland University of Technology completed in 1990

He has been a high school teacher in Math, Physics and Chemistry for 33 years.

He is the Head of the Science Department at Redcliffe State High School, Redcliff, Queensland, Australia. He has held this position since 1991.

Tony has been teaching Robotics Technology as part of the school's Physics course since 2002. In that time Tony has presented workshops in robotics and data logging throughout Australia.

He received a Westfield Scholarship in 2006 to study Robotics Education in the United States at Tufts University and at the Carnegie Mellon Robotics Academy. It was here he met the coauthors and fellow physics teachers Bill Church and Natasha Perova. They worked together to present at the first LEGO Engineering International Conference in 2007 and collaborated for this book to produce practical activities for teachers of Physics using LEGO technology.

Tony has a vision to see science education integrated with engineering through technology applications such as LEGO robotics. He has implemented technology-based courses at his school where students have a valid preparation and pathway to a career in physical sciences and engineering. Here he has seen more than 70% of his students, who have completed these courses at the high school level, continue to tertiary studies in these areas. At the junior high school level, all students have education in robotics programs as part of their science course and this has motivated enough students for two full elective physics and engineering classes in year 10, from a cohort of 200 at a public high school.

Tony is dedicated to the project and inquiry based approaches to physics that leads to motivated students, particularly boys who engage well with both a hands on and background theory approach. This book is part of that vision to see students enjoying physics with the latest technology.

Natasha Perova

Natasha received her M.S. in Mathematics, Science, Technology and Engineering education in 2008 and M.S. in Electrical Engineering in 2005 from Tufts University and B.S. in Electrical Engineering from Suffolk University.

Natasha is currently working at the Harvard Graduate School of Education as a research assistant focusing on students' learning algebra. She also teaches an introductory physics course for Suffolk University.

Natasha is very interested in finding ways to engage students in meaningful learning and sees LEGO robotics with data logging capabilities as an opportunity to teach students physics concepts in an exploratory and exciting way. By providing students with some ideas for projects in this book, we hope to provide a framework for learning where students will engage in scientific discussions, develop technical literacy, and most importantly have a chance to test out their ideas in creative ways.

Acknowledgements

I would to thank Natasha and Tony for writing this book with me. The best part about writing a book of activities with two other authors is that I have tripled the ideas I now have to use in my own classroom! Their expertise, perspective, creativity, and willingness to deal with the 14-hour time difference between the eastern United States and Australia kept this project moving! This book would not have been possible without the input, brainstorming, and activity testing from the students and teachers of the Littleton school district in Littleton, New Hampshire as well as the students and staff from the Center for Engineering Education and Outreach at Tufts University. In particular, I want to thank my principal Alan Smith, and my collaborators on science, robotics, and other educational technology projects: Chris Rogers, Brian Gravel, Adam Carberry, and Melissa Pickering. Their support and input have provided me with the inspiration, tools, and research background to give students many different approaches to the challenging and important task of learning science and engineering. Many thanks go to Steve Roberts, Paul Williams, and Lori Langlois for inviting me to join their professional development team, helping me to improve my teaching craft, and sharing my physics activities with an excellent community of New Hampshire teachers. I also want to thank Barbara Bratzel, Eric Wang, Bill Kearns, Claude Baumann, David Vernier, Vince Lunetta and our publisher Jim Dally for critical input on the chapters of this book. Finally, I want to thank my amazing wife Sarah and my daughters, Jessie and Theo, for their enthusiasm and willingness to chase robots around the house and classroom. This project would not have happened without their help, support, and guidance!

William Church

For feedback to Investigation activities, I want to thank the students at Redcliffe State High School, Queensland, Australia. For information related to the Physics Open Investigations and Criteria Sheets, I want to thank Glenn Rossiter and physics teachers in Education Queensland.

I would also like to acknowledge my lovely wife, Sue, who served me copious cups of Tea while I worked and because her English is so much better than mine, keeping me on track. Above all others thanks to my fellow authors for their dedication to children and learning and their perseverance with a difficult international collaboration as we "Skyped" late into the night or before dawn depending on where you lived.

Tony

I would like to thank my collaborators, Bill and Tony. It was great fun working with them and I have learned so much from their experiences in innovative teaching. I want to thank the National Science Foundation for funding part of my masters, work that got many of my ideas in this book started.

Special thanks to the Tufts CEEO where I worked for 2 and a half years. Graduate students and staff at the center are driven by the mission to improve education through engineering. By developing education tools, doing outreach and educational research, they are greatly contributing to increasing students' excitement for learning science, technology, engineering, and mathematics.

For all of the creative ideas about teaching and learning that helped me in developing the activities, I would like to thank Robert Hart, Chris Rogers, Barbara Bratzel, Judah Schwartz, Arthur Eisenkraft, Mike Barnett, Jon Star, David Henry Feldman, Joseph Noonan, Walter Johnson, and finally my parents for their patience and support.

Natasha

This book is based upon work supported by the National Science Foundation under Grant No. 0307656. Any opinions, findings, and conclusions or recommendations expressed in this material are those of the author(s) and do not necessarily reflect the views of the National Science Foundation.

Table of Contents

Chapter 1

Overview

1.1 Introduction

Why physics with robotics? Two reasons – inherent to robotics is measurement and actuation. In other words, in order for a robot to interact with the world, it has to investigate and do something with that investigation. It has to apply what it has "learned". Therefore, we can say that inherent to robotics is investigation and application. The same is true for physics education. In classrooms around the world, we encourage our students to investigate and apply what they have learned to something in the real world. Robotics gives us these two aspects of physics education in one very well connected package. Throughout this book, you will learn about opportunities to use robotics in your classroom for investigation and applied physics projects.

These are not, however, the only reasons we use robotics in our classrooms. Robotics is engaging, promotes creative problem solving, encourages students to represent their ideas in the real world with a precision system, provides excellent feedback and finally, robotics can be easily connected to knowing and learning a great diversity of physics concepts and skills.

Why LEGO® MINDSTORMS® based robotics? Combined, we have more than 15 years of experience with this system. While we have used other systems in physics instruction, we have use the MINDSTORMS system the most. We have found it to be an affordable, durable, and flexible solution for our project needs. Our experience with these systems over many years has allowed us to develop a diversity of activities and test them in our own classroom. However, if you are reading this book and have experience with fischertechnik® systems or those based on the BASIC Stamp®, PIC, Cricket, Arduino, or other microprocessors, try them out. The core of our ideas will work with most any sensor/motor based robotics system.

Whether it is LEGO based or not, robotics is a great hook in a classroom environment, engaging reluctant and enthusiastic learners. If your experience is similar to our own then you will see that the instant you take out the robot kits, the students are on the edge of their seat, waiting to dig in. You will see that these systems provide students with exceptional feedback, helping them to develop their ideas and successfully tinker with physical phenomena as they endeavor to take on various science and engineering challenges. Furthermore, you will see that after class has ended, you still have students in the room. They do not want to leave. They are having a great time challenging themselves and they are often so proud of what they have made that they bring friends later in the day to see their creations.

1.2 Supplement to Your Existing Curriculum

The purpose of this book is not to replace your current physics curriculum. Instead, the purpose of this book is to supplement your curriculum. Being teachers ourselves, we are motivated to share with you a set of activities centered on LEGO robotics because they work well to engage our own students in the process of learning the concepts and skills of physics. But, being teachers, we know that every school and every class within every school is different and those differences warrant adaption. In writing this book, we wanted to focus on providing you with a resource from which you can draw as you utilize LEGO MINDSTORMS in your own classroom. Therefore, while the activities of this book do not require any modification on your part to use in the classroom, we encourage you and your students to do so. That is, after all, how we created this book. We supplement our own physics classes with new ideas we discover each year. The ideas that inspired the activities of this book often came from conversations we had with our own students, fellow teachers, scientists, engineers, and other specialists in fields related to physics education. We hope that the ideas you find in this book inspire you and your students to create unique and powerful learning opportunities in your classroom.

1.3 LEGO® as a Physics Learning Technology

The goal of this section is to describe how the tools of a LEGO MINDSTORMS robotics kit fits into a physics learning environment. Because the book itself serves to demonstrate specifically how the LEGO MINDSTORMS kit is used in a physics class, our approach in this section is to treat the topic more generally, explaining more about why we chose these tools in our own physics classrooms. This section is also designed to be “food for thought” for you as you start to modify our activities and invent your own.

Generally speaking, the components of a LEGO MINDSTORMS kit serve as both measurement and design tools. As a measurement tool, they are only limited by the type of sensor to which you can connect. Both the NXT and RCX can be connected to many LEGO and non-LEGO based sensors. As a design tool, they afford the student a great deal of precision in their measurements and they allow students to use the same medium for designing scientific experiments as well as solutions to engineering design challenges. What follows is a more detailed discussion of these main points.

LEGO® MINDSTORMS® as a measurement tool

With LEGO built light sensors, thermometers, angle sensors, microphones, and distance sensors, both the RCX and NXT serve as stand alone measurement tools. Both the NXT and the RCX have a "view" option that allows you to see the data streaming in on any sensor directly on the device's screen (see Figure 1.3.1). Figure 1.3.2 provides two NXT based examples. The NXT can be set-up quickly for a diversity of short and simple investigations.

With a little more effort, the NXT can be extended to display data from third party sensors as well. For example, Figure 1.3.3 shows a connection between the NXT and sensors made by Vernier Software and Technology. Used this way, one can turn the NXT into a customizable data display center.

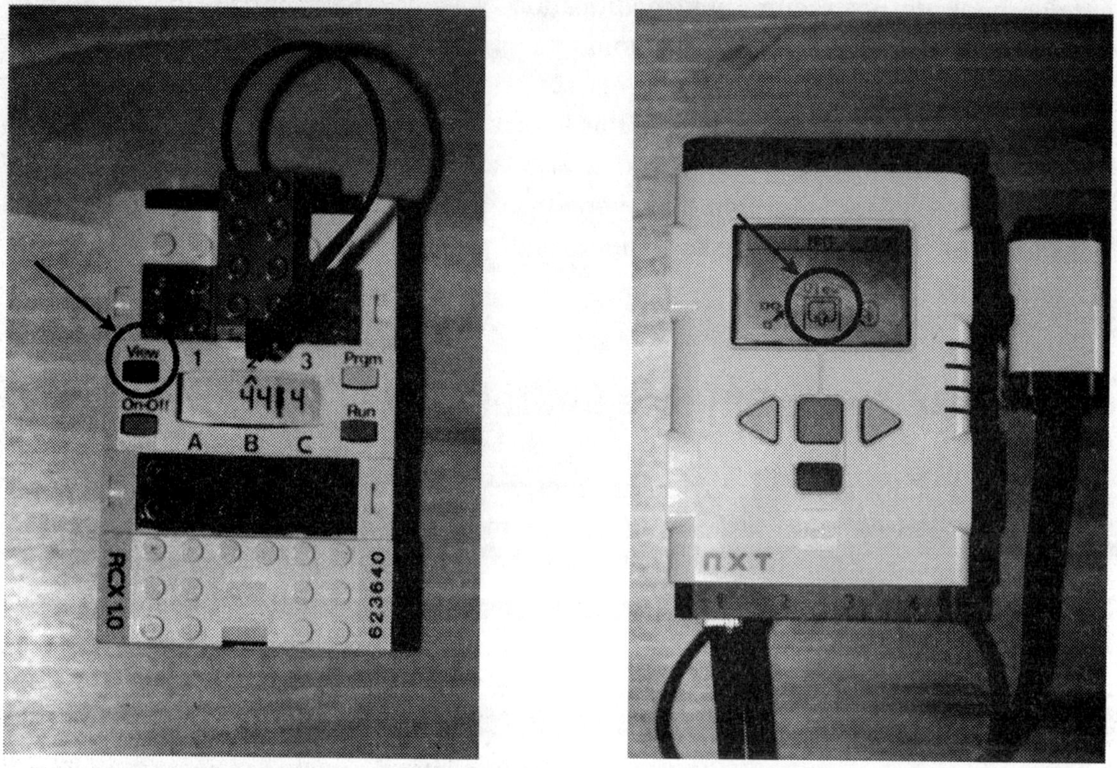

Figure 1.3.1 Use the "view" feature to display data from a sensor.

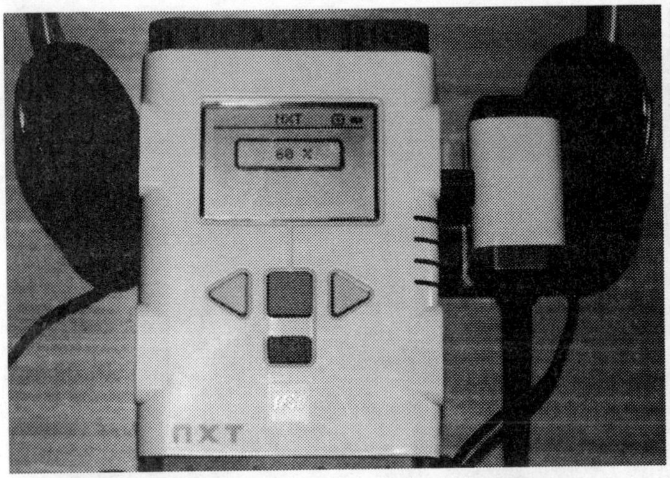

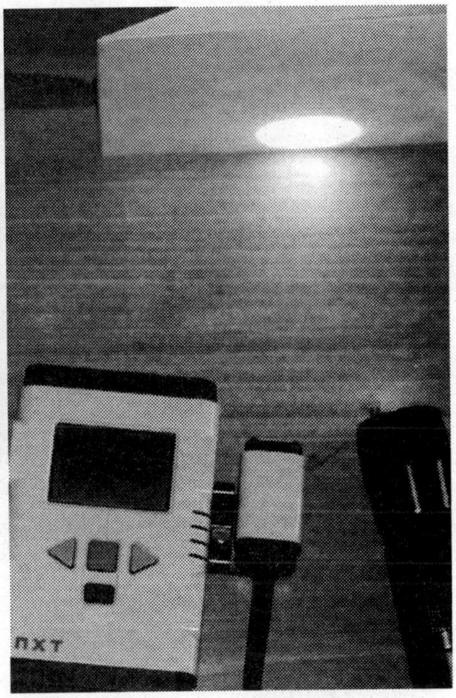

Figure 1.3.2 Use the NXT as a stand-alone measurement device.

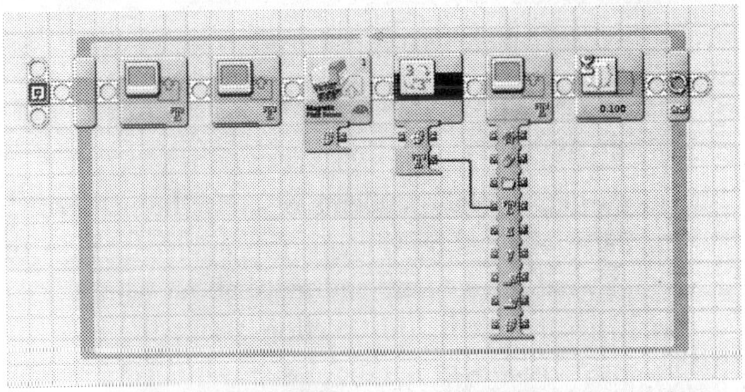

Figure 1.3.3 The NXT will work with 3rd party sensors. The graphic to the right shows the short MINDSTORMS code required to display data from Vernier sensors.

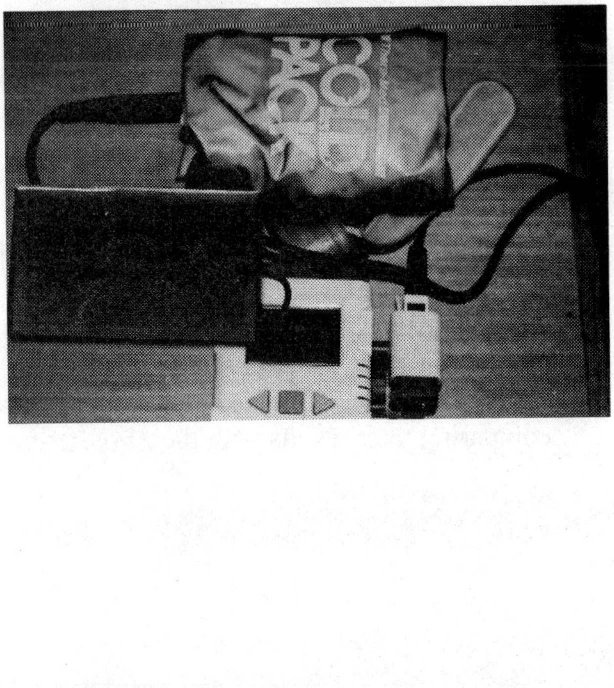

Fig. 1.3.3a Temperature sensor.

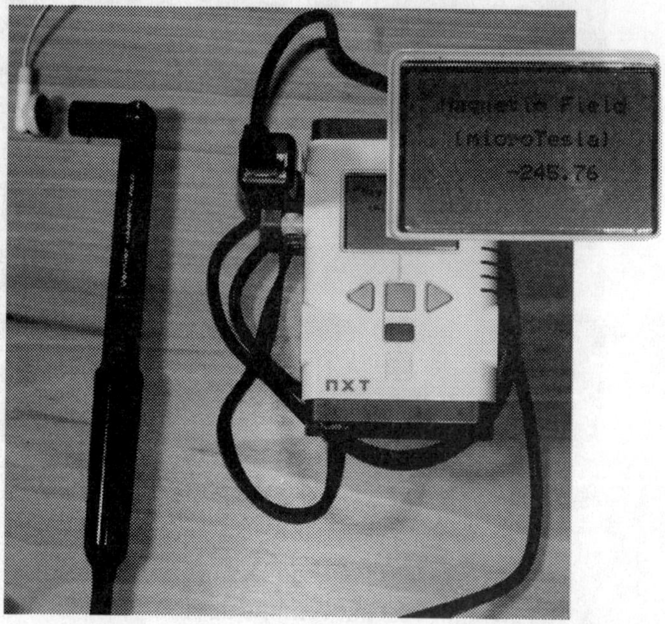

Fig. 13.3b Sound and magnetic field sensors.

http://www.vernier.com/nxt/downloads.html is the URL for downloading the Vernier sensor_block.

Using LEGO MINDSTORMS for accuracy and precision in your investigation and design activities

How often do your students discuss "human error" in their lab reports? How often are the data collected by students inconclusive? How often are discussions with students less about what the data are and more about what the data should be. The final point that we wish to discuss in this section about the role LEGO MINDSTORMS plays in your physics classroom is the issue of accuracy and precision in robot based physics investigations.

While the robotic tools used in the activities of this book enhance student investigations and applied physics design projects, they do not take the thinking out of the process. The accuracy of a student's measurements communicates whether or not the student measured what she wanted to measure. LEGO MINDSTORMS do not necessarily make a student's measurements more accurate. Your students will still need to learn and practice the essential skills of experimental design and the methods of science. They will need to determine the parameters of their experiment such as which variables to control, instruments to use, units of measurement, the duration of each run, and the number of runs. They will practice how to analyze data, reading trends and fitting appropriate models. They will practice evaluating the results of an investigation and comparing their results with their predictions.

LEGO MINDSTORMS is very helpful in improving the precision of student measurements. The precision of a set of measurements communicates how close the measurements are to each other. Precise measurements have a very small variance. NXT and RCX based experiments come in quite handy when precision is needed. If you program the robot to move for 5 wheel rotations, it will travel the same distance every single time[1]. If you program the robot to move its motor at a specified speed, it will do that every single time. Figure 1.3.4 is an example of the precise position vs. time data possible with an NXT based robot.

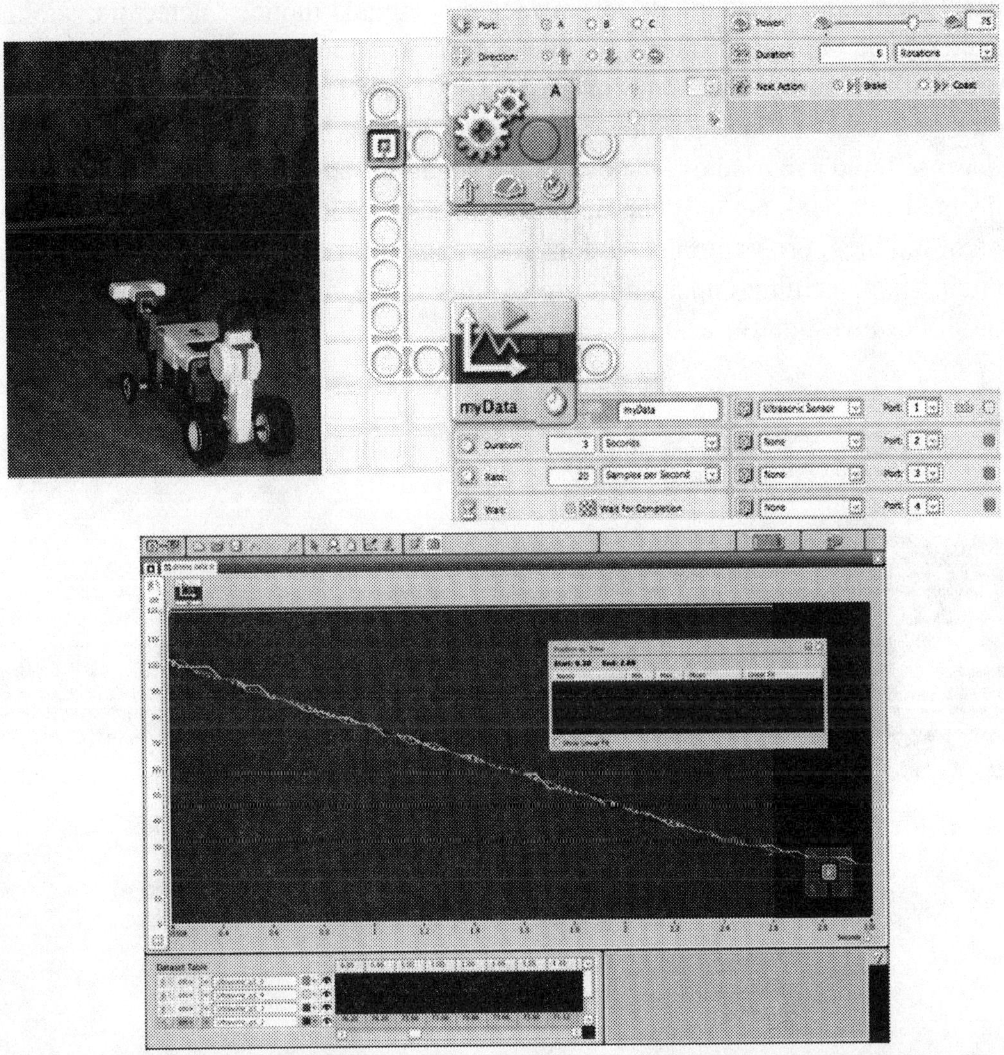

Fig. 1.3.4 Distance versus time measurements for a robot approaching a wall using LEGO MINDSTORMS Education NXT Software (version 2.0).

---

[1] Note: depending upon the floor surface, wheel choice, and whether the robot was programmed to coast to a stop or stop abruptly, the actual distance travel may not be the desired "5 rotations". Thus, without good experimental design, the measurement may not be "accurate". What will happen, however, is that the robot will travel the same distance each time. The set of measurements will be precise.

With the exception of low battery issues, you and your students will be very pleased with the precision of the data collected. This will be evident in the activities of this book. Experiencing the precision of data in a LEGO MINDSTORMS based investigation opens up your mind to investigations that you did not think possible before.

LEGO MINDSTORMS as a physics investigation and design tool

As a measurement tool, the NXT and RCX stand with meter sticks, stop watches, spring scales, thermometers, digital force meters, digital motion detectors, voltmeters, ammeters, and other instruments in your physics equipment inventory. But, what about the rest of the pieces that come in a LEGO MINDSTORMS kit? LEGO is, after all, known as a construction medium. The LEGO blocks, beams, wheels, gears, etc allow students to build a myriad of set-ups for investigations and applied physics projects. The projects of this book are only the tip of the iceberg. Once you are comfortable with the ways of building, programming, and data logging with LEGO MINDSTORMS, you will see many new learning opportunities to explore in the future. See Figure 1.3.5 for two examples of investigative design.

Figure 1.3.5 Limitless investigation design possibilities exist. Shown here are examples of an incline plane with reduced friction and a physical pendulum for studying the period of a vibration.

Bridging the gap between physics and engineering

Engineering design challenges are not new to physics classes. From paper airplanes, egg drops, water rockets, toothpick bridges and mousetrap cars, engineering challenges in physics instruction allow students the opportunity to engage in creative, enjoyable, and practical ways. Engineering design challenges gives students an opportunity to talk about physics as it relates to something they created, something practical. Engineering design challenges puts physics to use and immediately answers the question, "Why are we learning this?"

If physics teachers already do engineering design challenges, how do LEGO MINDSTORMS kits enhance this form of instruction? They maximize the ratio of equipment to project possibilities. With egg drops, toothpick bridges and mousetrap cars, you need to obtain and maintain a steady supply of materials, each set of materials dedicated to only a few types of projects. With LEGO MINDSTORMS the number of projects is almost limitless. Perhaps you will not do an egg drop project with MINDSTORMS or launch an NXT in a rocket[2], but with one kit of materials, you can do many other very engaging applied physics projects.

In this book, we emphasize the link between physics investigations and engineering design by providing activities that show students the need to investigate *while* taking on an engineering design challenge. For example, students will investigate gear ratios while creating a motorized crane or drag car and investigate sound waves while creating a system to make the best ear protection. By having students engaged in projects that synthesize investigations with engineering design, you are helping them close the gap between the concepts and skills of physics and the practical use of those skills. In closing the gap, we help students take what they learn in the classroom and use it in the rest of their lives.

1.4 Companion Website

Please visit http://www.physicswithrobotics.com for additional resources related to the content of the book. Here, we will provide downloadable content such as programming, building, and other activity resources. We will also provide a reader feedback system to allow you to share your classroom experiences related to the book and to help us improve future versions of these activities.

[2] Visit Hhttp://www.unr.edu/nevadasat/HALEH for a neat example of MINDSTORMS in near space!

References

Bratzel, B. (2007). *Physics by Design, 2nd Edition, ROBOLAB Activities for the NXT and RCX.* Knoxville, TN: College House Enterprises, LLC.

Erwin, B. (2001). *Creative Projects With LEGO® MINDSTORMS®*. Boston, MA: Addison-Wesley.

Wang, E. (2007). *Engineering With LEGO® Bricks and ROBOLAB™, 3rd Edition.* Knoxville, TN: College House Enterprises, LLC.

Chapter 2

Getting Started: Part I. **Hardware.**

2.1 Introduction

The purpose of Chapters Two and Three are to introduce you to the hardware and software tools that will allow you to begin to use LEGO® MINDSTORMS® in your physics class. Chapter Two focuses on hardware, providing you with background information, tips, and techniques to get started with the sensors, motors, beams, gears, and two different types of LEGO vehicles used in many activities of this book. Chapter Three focuses on software, introducing you to several software options available for programming and data logging with the NXT or RCX.

As "getting started" chapters, these chapters are not a complete resource, answering all of your hardware and software questions. Instead, they are a starting point helping you to begin building, programming, and operating NXT or RCX based physics models while collecting data. If you are new to using LEGO based robotics in your classroom, this chapter will help you decide what to obtain and how to begin. If you are experienced, this chapter will inform you regarding hardware and software options that you may not have explored yet. For topics not covered in the book, refer to the reference section at the end of this chapter for pointers to the many exceptional resources available in print and online.

Please note, as with many computer-based technologies, the tools we present in this book have very high ceilings. In other words, you and your students can take them to moon if you wish. This is good and bad. On the good side, it means that you probably have a solution in hand to your most complex investigation or engineering based physics projects. On the bad side, it means that there is the risk of getting lost in the technology, missing the point of a given investigation or engineering task. Using Albert Einstein's words of wisdom as a guide -- "Everything should be made as simple as possible, but no simpler" -- we present to you a chapter that introduces the hardware tools for exploring physics with robotics.

2.2 Comparing your Hardware and Software Options

If you are at the beginning of the process of choosing which hardware and software you will use in your classroom, you have many options. To reiterate a point that we made in Chapter 1, we do not present all of the options for using robotics in a physics classroom but rather the options that we have thoroughly explored in our own classrooms. If you would like to learn about more options, visit http://en.wikipedia.org/wiki/Lego_MINDSTORMS for a fairly thorough treatment of the topic.

As for hardware choices, Figure 2.2.1 represents a comparison between the NXT and RCX. While features are important, we also understand that price and availability are also important. Explore resources within your own school district. You may be surprised to find a supply of kits that you can borrow. Also consider buying used. 2009 is the last year that LEGO will be selling the RCX. You may find schools or individuals looking to transition from the RCX to the NXT, giving you a good deal on older RCX hardware. This platform is more than adequate for many of the activities found in this book.

	RCX with ROBOLAB (RL) 2.9	**NXT**
Number of Motor ports	3	3
Number of Sensor ports	3	4
Rechargeable Battery System	No	Yes
IR Compatible	Yes	No
Bluetooth Compatible	No	Yes
Multiple Brick Communication	Yes	Yes
Motor speeds	100	100
Sounds	6 beeps	infinite
Drive straight	No	Yes
On-board variables	64	100-1000
Display control	numbers	text, images, numbers
Number of stored programs	5	infinite
On-brick programming	No	Yes
Uses new and old sensors	(No)*	Yes
Hard-wire link to computer	No	Yes
3rd party sensors	Tough	Easy
On-board graph display	No	Yes
*You can use the NXT Touch and Light sensors in RL 2.5.4 and all 4 NXT sensors + servo motor in RL 2.9		

Figure 2.2.1 Table comparing the features of the NXT and RCX microcontrollers. Table used with permission from LEGOEngineering.com.

As for software choices, Figure 2.2.2 represents the software choices covered in Chapter 3 of this book. One option missing from this table is LabVIEW™. Instead of treating LabVIEW separately, we demonstrate in Chapter 3 how to access LabVIEW features through the ROBOLAB product. If you would like to use LabVIEW for your projects and need more information, look at the resource section at the end of this chapter.

	Software for NXT	Software for RCX
Programming	MINDSTORMS NXT Software 1.0-2.0 Robolab 2.9	Robolab 2.5 - 2.9
Data logging	MINDSTORMS NXT Software 2.0 Robolab 2.9 Logger Pro 3	Robolab 2.5 - 2.9 Logger Pro 3

Figure 2.2.2 Programming and Data Logging software options for NXT and RCX

2.3 NXT Motor and Sensors

In the following section, we wish to take a closer look at the sensors and motors that are included in a LEGO® MINDSTORMS® Education NXT Base Set. For each of the kit's sensors and motors, we outline how they work and which physics concepts can be connected to them most easily. Knowing how these components work and how they connect with physics concepts will help you facilitate and troubleshoot the activities of this book. For instance, knowing that the distance sensor works via reflections of a sound wave will help you and your students know from which surfaces you can and cannot obtain distance readings. Knowing that the light sensor can be operated with its LED on or off will help you determine if the light readings you have collected are from the ambient light of the room or from the reflected light of the sensor's LED. While we have chosen to focus on the NXT kit, the information we provide can be generally applied to RCX's sensors and motors. A few differences are worth noting here:

- The RCX motor does not contain a rotation sensor but you can integrate one into your projects using LEGO's rotation sensor (LEGO Part #9756).
- LEGO does not make a distance sensor for the RCX.
- LEGO does not make a sound sensor for the RCX.

If you need either a distance sensor or a sound sensor for the RCX, start your research at www.extremenxt.com/lego.htm.

To illustrate how the concepts apply to the NXT's sensors and motor, we also provide you with several investigation and design ideas. The main goal of providing the project ideas here is to clearly illustrate the concept connections. Not all of the project ideas included here are tied to the activities of the book. For those ideas that are not tied to the activities of the book, consider them as starting points for open ended activities.

NXT Interactive Servo Motor

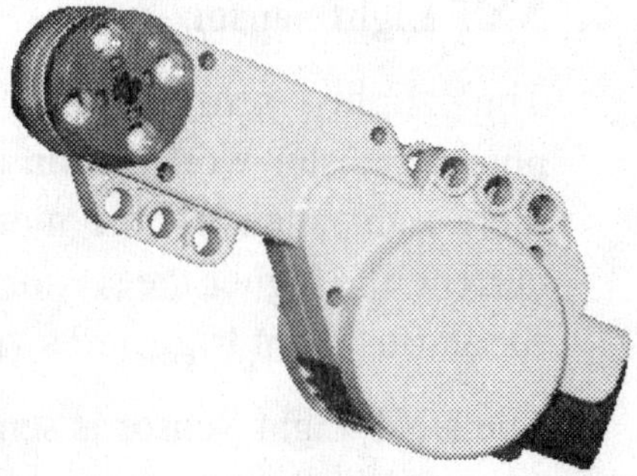

Servomotors are a combination of an electric motor, position sensing device and a feedback circuit to control the motor. They are often used in radio-controlled devices such as model boats and planes. Pulse Width Modulation (PWM) controls servomotors where current is generated for certain intervals and at a rate of about 20hz. PWM is a fast switch, turning the motor on and off. If the motor is operated at a 50% power level then the PWM will cycle the motor on 50% of the time. If the motor is operated at 10% power then it will be cycled on 10% of the time and cycled off 90% of the time. PWM is a convenient way to enable software commands to control motor speed.

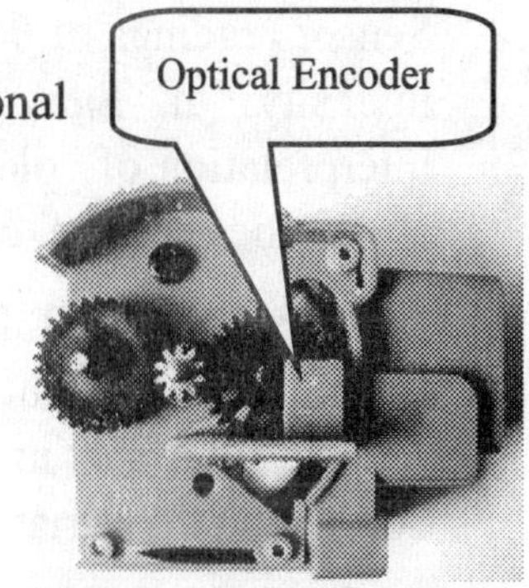

Photo used with permission from Philippe Hurbain

The NXT motor is quite complex compared to the traditional 9V LEGO motor with the RCX kits. It has a higher ratio gear train and an optical encoder that reads the rotations of a slotted wheel. This encoder is accurate to the nearest whole degree. The gear train will make the motor slower than the traditional motor but produce more torque. Being a servomotor it is very useful for precision movements like robotic arms and positioning robots. It is very sturdy compared to the old motors. It is much less likely to be damaged while stalling and has no spindle that can snap off.

Concept Connections

Velocity
Acceleration
Force
Torque
Frequency
Electromagnetism
Circuits
Power
Electric current

Project Ideas

- Investigate simple parallel and series circuits with a motor and a 9 volt battery
- Add external gears to the motor and measure the torque produced
- Measure speed of a LEGO vehicle, predict times and distances

Web References

http://www.philohome.com/nxtmotor/nxtmotor.htm
http://www.philohome.com/motors/motorcomp.htm

NXT Light Sensor

The light sensor uses a phototransistor. A phototransistor works as an amplifier in a circuit. The more light intensity, the more current gain. The more current, the higher the sensor readings. These are useful for alarms, light triggered switches, and control circuits.

The NXT Light Sensor is sensitive to changes in light intensity instead of frequency or wavelength (i.e. color). While in a strict sense, it does not measure wavelength, its sensitivity does depend upon the wavelength of incoming light. For the NXT light sensor, the sensitivity peaks at a wavelength of 860 nm, which is in infrared portion of the spectrum. With a peak in the infrared, this sensor will yield readings for *both* visible and infrared sources.

Sensor readings of 0-100 represent dark to very bright. Like your eye, it detects the intensity of light with essentially monochromatic vision. Some minimal interpretation of color might be done by the degree of light reflection but it won't always be accurate, especially if the incident light is not white.

Built into the NXT Light Sensor circuit is a red LED that can be turned on or off (using ambient light only). This can aid in detecting changes in brightness of a reflected surface in darker rooms. Light sensors are not good object detectors because of the variation in object color and ambient light levels.

Concept Connections

Light Waves
Color
Light Intensity
Light Reflection
Light Transmission
Light Emitting Diodes
Transistors
Frequency

Project Ideas

- Measure the intensity of a light source as a function of distance
- Measure the reflectivity of various surfaces as a function of color
- Measure the turbidity of water.

Web Reference

- http://www.engr.udayton.edu/faculty/jloomis/ece445/topics/egginc/tp4.html
- http://MINDSTORMS.lego.com/Overview/nxtreme.aspx
  Download the "Hardware Developer Kit"
- http://www.arrownac.com/
  Search for part # SFH 309-4

NXT Touch Sensor

The Touch sensor is a single pole, double throw switch behind a spring that toggles on and off after each press. It is referred to as a digital sensor because it only has two states.

The NXT touch sensor can be used for control programs such as a touch sensitive burglar alarm; a hand operated remote control; and starting and stopping motors or sensor readings.

Concept Connections	Newton's 3rd Law Simple Circuits Reaction times
Project Ideas	• Test Newton's third law by using the touch sensor on a moving object vs. a fixed object. Which triggers the sensor? • Use the touch sensor in a reaction time experiment. Program the NXT to turn on a light and start a timer. When the switch is activated, the timer stops. • Using only a 9volt battery, NXT to RCX cable, touch sensor and motor, build a circuit to turn on and off the motor with the touch sensor.

NXT Sound Sensor

Sound level sensors are used to represent the loudness (or amplitude) of a sound wave rather than its frequency. Frequency measurements require very high sampling rates while amplitude measurements do not. While the NXT does not have an adequate sampling rate to measure sound frequency, it can measure sound amplitude. The sound level sensor works by measuring changes in air pressure. The change in pressure on the diaphragm inside the sensor is used to measure the *relative intensity* of the sound wave in units of decibels (dB). The range of typical relative intensity scale is from the threshold of hearing (0 dB) to the threshold of pain (120 dB). The relative intensity scale is a logarithmic scale that serves to reduce the enormous range of sound intensities (measured in *Watts/m*^{2}, for example) between the threshold of hearing and the threshold of pain.

By making the connection between relative intensity and intensity for students, you can help them understand that a sound sensor is really a sound wave *energy* sensor.

The equation for intensity is:

$$\text{Intensity} = \frac{\text{Power}}{\text{Area}}$$

By replacing power with,

$$\text{Power} = \frac{\text{Energy}}{\text{Time}}$$

the equation for intensity can be rewritten as:

$$\text{Intensity} = \frac{\text{Energy}}{\text{Time} \times \text{Area}}$$

Here, we are relating the measurements derived from a sound sensor to the energy of the sound wave. While our assertions represent general relationships and not specific quantities, they are useful for a student trying to understand what sound level measurements mean and what factors affect sound level. By explicitly bringing energy into the discussion about sound level readings, it helps students connect their sound measurements with other aspects of their physics study.

This is especially important to consider when using the NXT sound sensor because the readings are not reported in decibels. Furthermore, sound sensor values vary according to the frequency of sound. When used in the dBA mode, the sound level sensor filters its response to more closely match the frequency to sound level sensitivity curve of the human ear. As an approximation, however, sensor values of 0 to 100 correspond to relative intensity readings of 50 to 90 dB.

Concept Connections

Relative Intensity vs. Intensity
Relationship between Intensity, Power, and Energy
Law of Reflection
Speed of sound

Project Ideas

- Investigate the relationship between distance and sound level
- Explore sound wave diffraction, reflection, diffraction and conduction in various materials
- Map large scale interference patterns
- Test the effectiveness of a string and cup set-up to transmit sound

Web References

http://www.convict.lu/htm/rob/NXT_sound_sensor.htm
http://www.brickshelf.com/cgi-bin/gallery.cgi?i=3348152
http://www.explainthatstuff.com/soundlevelmeters.html
http://hyperphysics.phy-astr.gsu.edu/hbase/ph4060

NXT Ultrasonic Sensor

Ultrasonic sensors determine the distance to a target by sending out sound waves at ultrasonic frequencies and listening for the returning echoes that are reflected from the target. The time (*t*) that it takes for the sound pulse to travel out to the object and then back to the sensor is the basis for obtaining the distance (*d*) measurement. Because this measured time is twice the value desired, the distance between the object and the sensor is:

$$d = v_{sound} \frac{t}{2}$$

The LEGO ultrasonic sensor's operation and function is almost identical to popular motion detectors used in many physics labs. A key difference is that the sensor has a separate transmitter and receiver (left and right eye), and can therefore be used to measure objects at closer range. Ultrasonic sensors that use the same diaphragm need to build a "wait" into the system so that the transmission vibration can dampen before being read as it resonates with the echo pulse. A comparison between the performance of LEGO's ultrasonic sensor and other "motion detectors" such as one from Vernier Software and Technology would be an excellent activity.

As stated in the product literature, the recommended range and accuracy of the sensor is 0 to 255cm +/- 3cm. If this level of performance is not being achieved, knowing the physics behind the sensor will help you troubleshoot the situation. For example, reflection of sonic waves can be just as diffuse as light from uneven surfaces. Therefore, detecting small, non-flat, is difficult. These objects are acoustically opaque. Even larger flat objects can give you trouble if their surface is not normal to the incident beam. One more issue related to the performance of the sensor is the temperature of the room. Although, the impact of variations in classroom temperature in the range of $^{+}$/-5 degrees is still within the expected error bars. It is a good challenge to give to students to have them figure out the range of temperature readings that will impact distance data reported by the sensor.

For data logging tasks, any experiment requiring distance can use this sensor although the accuracy is likely to be less than that of the rotation sensor. However, in some investigation activities, the convenience of collecting linear distance without having to convert rotations into a linear distance is worth the sacrifice in accuracy.

For programming tasks, the distance sensor can be used for object detection and position control.

Concept Connections
Accuracy vs. Precision
Sound frequency range – ultrasonic
Law of Reflection
Reflectance vs. absorption
Speed of sound

Project Ideas
- Investigate which shape objects can be used with this sensor.
- Investigate the range of angles for a flat plan that work with this sensor.
- Investigate the types of materials that will absorb sound vs. reflect sound.
- Determine the range of reliable measurements. Test hypothesis to determine the essential variables to maximize the range.

Web References
http://MINDSTORMS.lego.com/overview/Ultrasonic_Sensor.aspx

2.4 LEGO Building Primer

There are many books and web resources that cover LEGO building techniques but a few tips might be handy. In this section, we present to you ideas focused on the following:

- Building Durable Structures
- Motors and Gears
- Example LEGO Vehicles (RCX & NXT)

Building Durable Structures

The strength of a structure comes from reducing stresses. Linear stresses are:

Tension – the stretching of supports.
Compression – the squeezing of supports.

Pieces of LEGO are very strong in these directions. When pieces are joined together, the joint can be strengthened with overlapping beams. If a torque is applied to a structure, it experiences a twisting stress. This will reveal weaknesses in a LEGO structure unless it is strengthened with ties and struts (see Figure 2.4.1).

As you design, you can reduce torque (i.e. the turning force) at any given point on a structure by reducing length to the pivot and/or sharing the weight along a number of beams.

Motors and Gears

The use of gears offers physics students an excellent opportunity to explore many concepts from mathematical ratios to speed, force, torque, work, and energy. Figures 2.4.2 – 2.4.6 represent design ideas for attaching gears to RCX and NXT based robot vehicles.

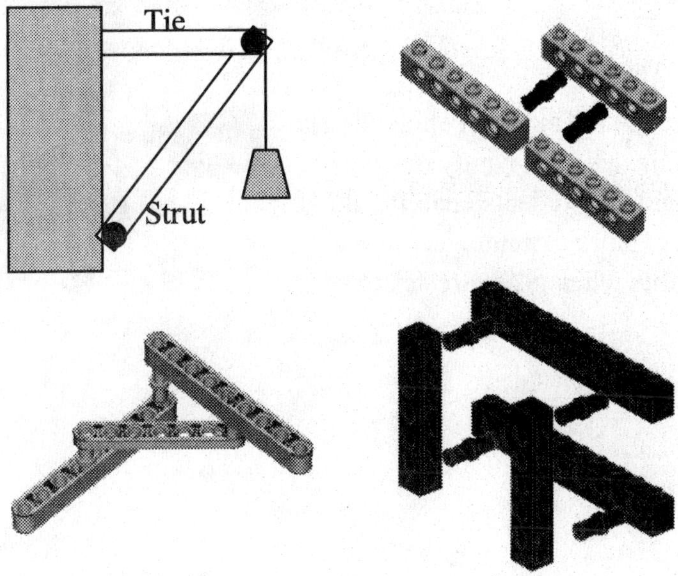

Figure 2.4.1 Examples of methods for strengthening joints

Figure 2.4.2 Gears on the RCX Motor. Mounting the right gears and connecting to your motor successfully is often tricky. There is a plate with a groove designed to do just that with the RCX motors. By joining two beams side by side with pins on this plate, you will have many holes to align your gears as well as provide a chassis with more support for your load and axles.

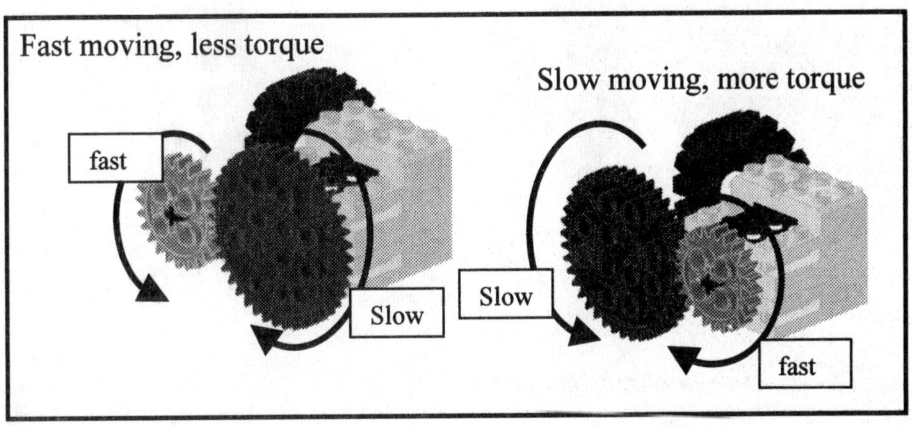

Figure 2.4.3 Gear ratios are very important to speed up your car or give it more push – sorry, you can not have both! This motor set-up utilizes the design represented in Figure 2.4.2.

Figure 2.4.4 This hand cranked gear train set-up is handy to have available to students enabling them to explore extreme gear ratios possible when gears are "stacked".

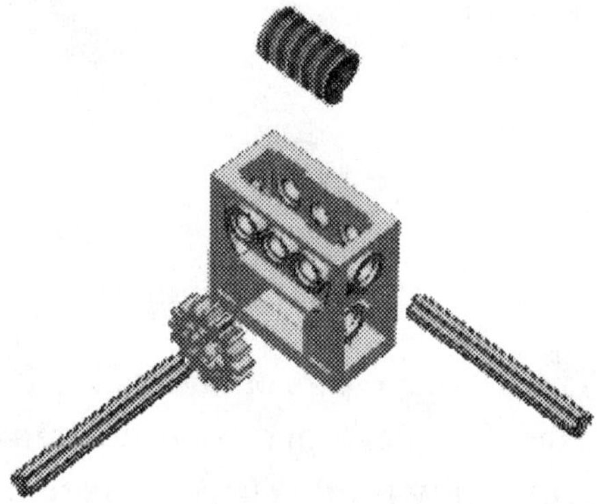

Figure 2.4.5 Strength. For super strong machines, use the worm drive on your motor.

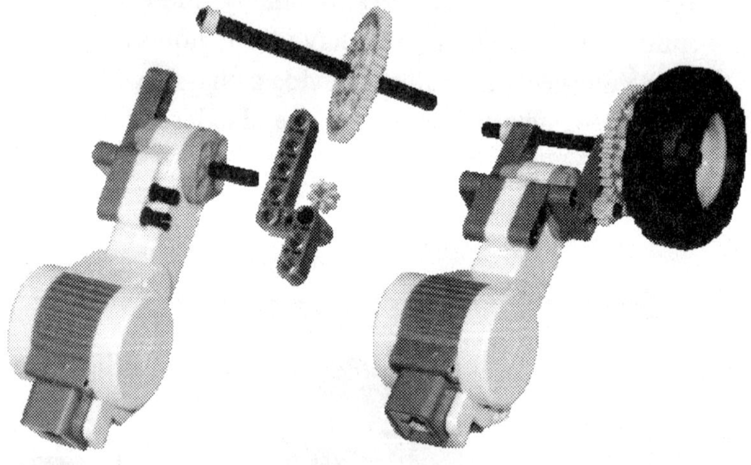

Figure 2.4.6 Gears on the NXT Motor. The NXT motor has built in gears that make it strong but slow. As shown here, you can slow it down even further. If you switch the motor and load gears, you can speed up the NXT motor.

Examples of Simple Cars you can build.

Figure 2.4.7 Single Motor NXT Car.

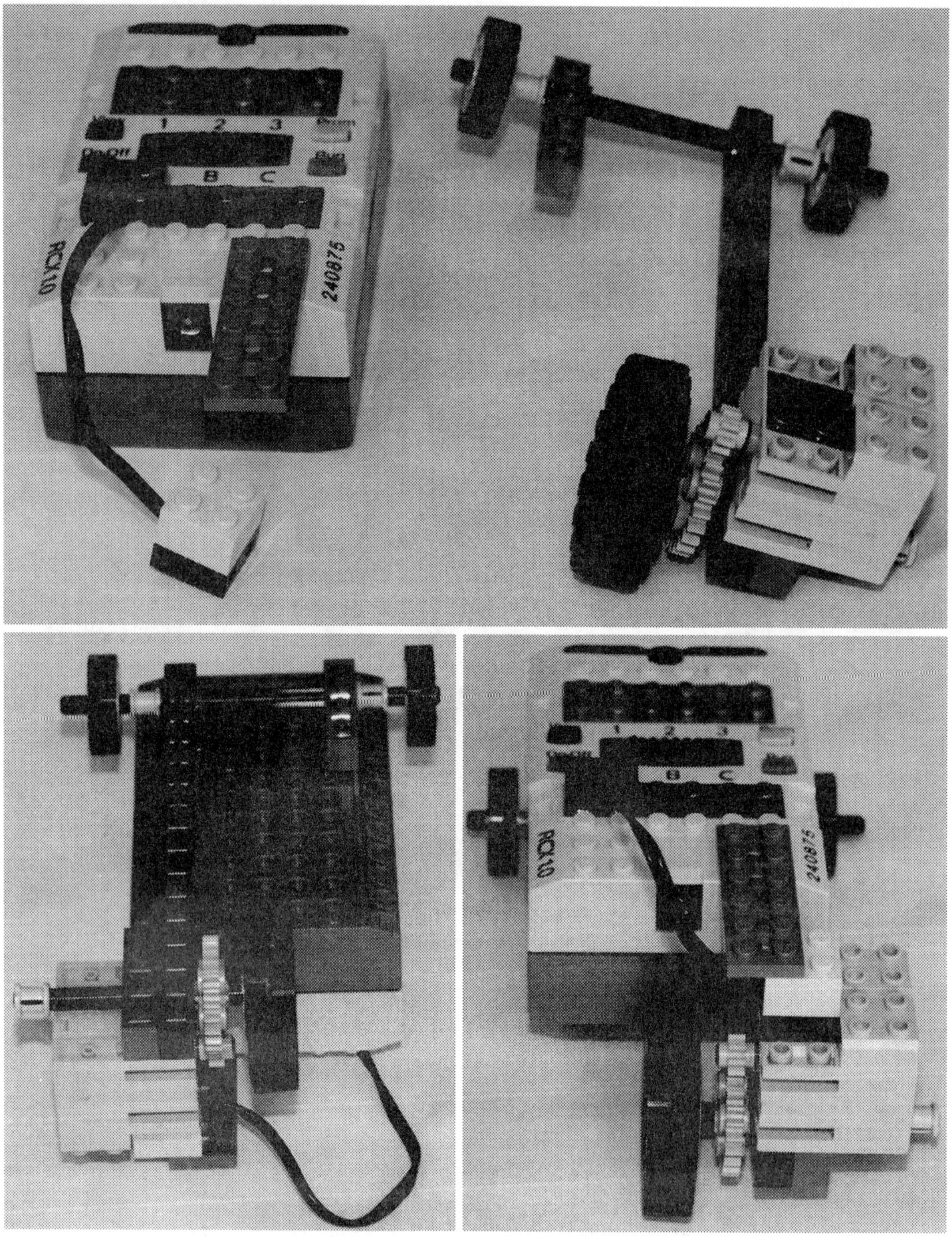

Figure 2.4.8 Single Motor RCX Car. The original design of this vehicle is from Eric Wang's book, <u>Engineering With LEGO Bricks and Robolab</u>, 2nd Edition.

Figure 2.4.9 Double motor NXT design from LEGOEngineering.com . Note, if you remove the rubber wheel from the front of the vehicle, it will turn more easily.

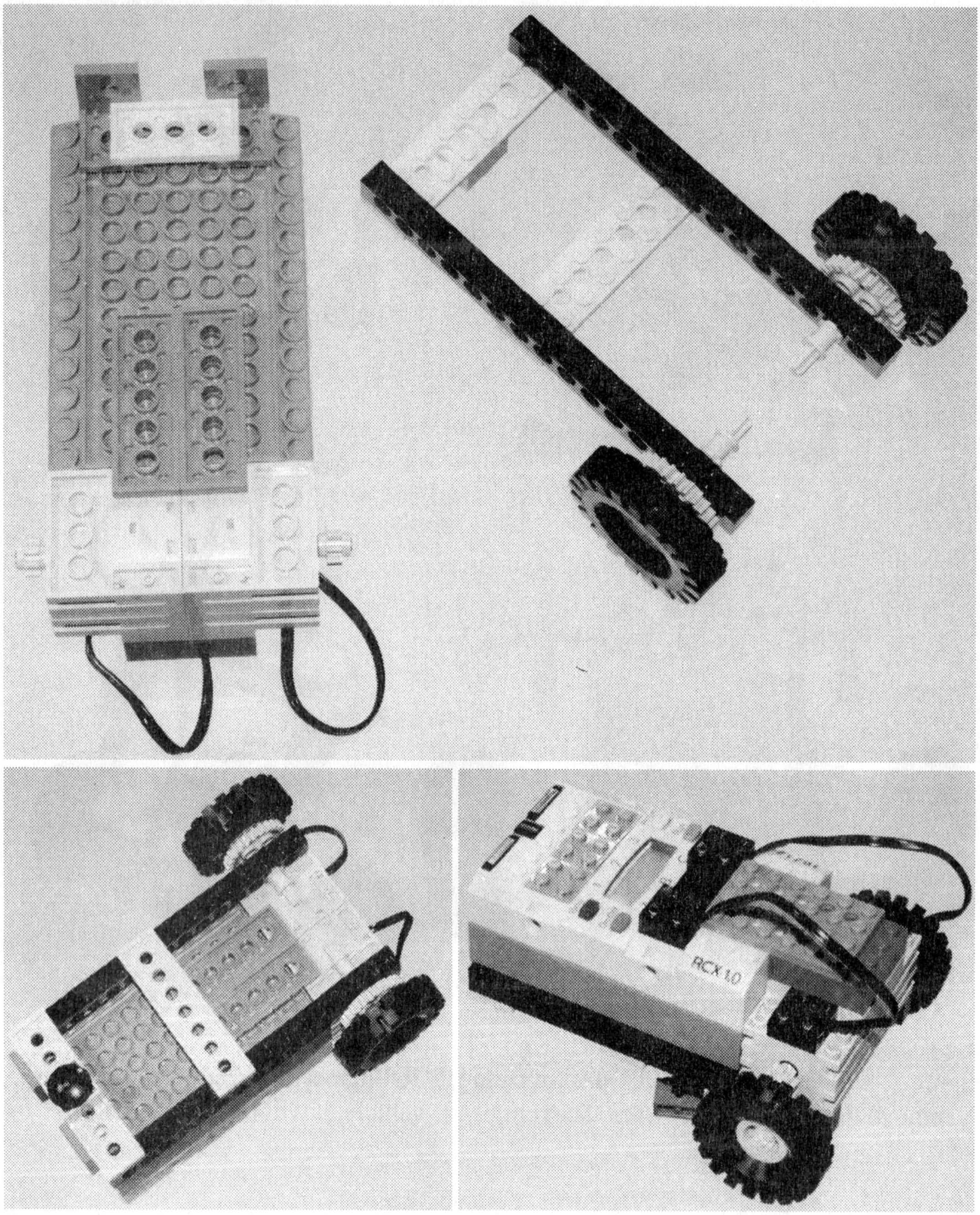

Figure 2.4.10 Double motor RCX design. The slider plate on the front of the vehicle enables the vehicle to turn easily without the complication of designing a caster type wheel.

2.5 Classroom Organization, Tips, and Techniques.

Room Organization

A significant challenge to overcome in LEGO activities is the organization of the equipment. This is particularly important if there are a number of classes sharing it. Recently, in one of the author's classrooms, students had just completed the Motorized Crane activity (Chapter 7) across eight classes. Three classes worked at the same time for three lessons each. This required the distribution and management of equipment for all three classes, including 24 RCX bricks and motors. It was successful and well worth doing but involved planning and effort by faculty and staff to have the gear in the right place at the right time.

When thinking about organizing your equipment, consider two options -- keeping the gear in separate kits similar in design to the original MINDSTORMS kits or sorting all of your equipment into boxes or trays. For small classes, keeping the pieces in kits can work well. It gives students ownership over the maintenance and inventory of their own kit of parts. If you need to keep a tight inventory of your parts, keeping organized and frequently inventoried kits is the way to go. If you have large classes or as in the example above, multiple classes, we recommend dividing all of the gear up into boxes or trays. Consider an organizational scheme that we use:

In a separate storage cabinet:

- Motors
- Sensors
- Wires
- Microcontrollers (NXT or RCX)

7 trays of Lego pieces:

- Blocks
- Beams
- Plates
- Gears
- Wheels
- Axles
- Connectors, bushings and other small pieces

Obviously, there are many other pieces but we use this organizational scheme to help students decide where to look for a part and to which tray to return a part. Consider this as a guiding principle: the easier it is for students to sort at the end of an activity, the more chance they put things away in their intended places. We tend to keep electronics and motors separate, locked away for safekeeping.

To facilitate the kind of movement common in robotics projects, we recommend an open style of room (Figure 2.5.1). This is a classroom where tables and cupboards are around the perimeter of the room and the center is left free for performance and investigation. You might use the floor or have a few tables, depending on what your doing.

Figure 2.5.1 Open style classroom with equipment trays and computer stations around the perimeter of the room.

General Classroom Tips and Techniques

What follows is our best attempt to present to you a diverse set of tips and techniques to help you and your students while completing the activities in this book. We recommend that you come back to this section after you have picked specific activities and even after you have completed these activities. Are we missing anything here? If so, visit our companion website (www.physicswithrobotics.com) and contribute your ideas!

To maximize the density of ideas, we present them to you in a bulleted list format:

Classroom and kit specific tips:

- An alternative system for organizing your kits is to reduce the number of elements per kit. For instance, if you anticipate that you will do several mechanics activities using the simple car designs we present in this chapter, create a kit that contains those pieces. The smaller number of parts will be easy for students to manage and easy to inventory after the activities are done. Extra pieces and add-on pieces can be kept as a whole class resource.
- Kits with limited pieces can also help students focus their design ideas, helping them achieve design goals quicker.
- If students are using the designs found in this chapter or other books, build at least one classroom model that students can reference as they construct their own models.
- Allow students to personalize their products. It may appear unproductive but it encourages ownership and helps students distinguish their project from others.

- If you already have lines on the floor such as tiles, use them to mark starts and finishes. Tiles are also a great resource for defining the goal of "driving straight". For example, several of the activities in Chapters 4 and 5 require vehicles that drive straight. Often, it is not necessary that the vehicles drive perfectly straight. Thus, we say to students, if your vehicle stays within this range of tiles while it drives forward for 3 meters then it is straight enough.
- Tape on the floor is a very useful resource for many robotics activities. Tape stuck on a tiled floor can last through a lot of wear and tear but check with your custodial staff and use a tape that does not interfere with their work. We recommend "painters" tape. This is a type of masking tape (often blue in color) that does not leave a gummy residue when it is pealed off the floor. Tape left on the floor for a long time can be scrapped up with a razor blade.
- Batteries, batteries, batteries! Create a system that works well for you and your students to help you manage battery charging. If you are using rechargeable AA batteries, set up a shelf or cabinet with separate bins for depleted and charged batteries. Write a date on your charged batteries if you know that it might be several weeks or more before they are used again. Investigate the properties of the batteries you choose so you can best accommodate for their deficiencies.
- AC Power. While both the NXT and RCX can be powered by an AC Adapter (LEGOEducation.com Product ID: W979833), this adapter will <u>not</u> charge batteries in your RCX. This adapter will only charge an NXT equipped with the NXT rechargeable battery pack (Product ID: W979798).
- Downloading firmware onto the RCX can be a pain. The RCX communicates with the computer via an infrared tower. Be aware of interference from other infrared sources such as classroom lights and sunlight. If you are having communication problems, make sure the RCX is on and that its batteries are good. Put the RCX and the tower in a dark spot. Be patient. Covering the RCX and tower with a box also works but make sure that you do not knock them out of alignment when you set the box down.

Data logging tips

- Check sensors to see that they meet your expectations by doing some initial readings.
- Explain typical graphs of sensor readings before students produce them. Often, students are put off by the spikes in a graph rather than the trend or shape. Collecting 25,50, or 100 readings per second produces "noise". Encourage your students to look for trends within the noise.

Team building and group processing tips

- Robotics projects work well as a group project. There are many tasks to complete; everyone in a group can make valuable contributions. Depending on what works best for your students, you have the option of defining student roles or encouraging the group to define group roles themselves. To encourage skill building, we recommend that students change their roles throughout the semester. Building, programming, data logging, and analysis work help students build a diversity of valuable science lab skills.
- Always leave a window before and after activities to brainstorm and review. For open style designs and investigations, students want a balance between freedom to choose and some guidance.
- Having students share what they did to solve a problem is a nonthreatening way for other students to learn and still feel they own their solution.

2.6 Resources

Companion website for the book:

- www.physicswithrobotics.com

Great ideas, resources and community building:

- www.legoengineering.com. A site developed by the Tufts Center for Engineering Education and Outreach and LEGO.
- www.MINDSTORMS.lego.com. The main LEGO MINDSTORMS site.
- www.vernier.com/nxt. Many excellent resources and project ideas related to the use of Vernier sensor with the NXT.
- www.education.rec.ri.cmu.edu/index.htm. The Carnegie Mellon University Robotics Academy.
- thenxtstep.blogspot.com. An up to date blog on all things NXT.
- www.brickshelf.com. An excellent and ever expanding resource!
- www.lugnet.com. The unofficial, international LEGO users group network.
- www.domabotics.com/index.php. Excellent classroom ideas and resources.
- www.convict.lu/Jeunes/RoboticsIntro.htm. Many advanced projects, resources, and ideas. An inspirational source for extended science and engineering projects. Some examples use Ultimate Robolab, the advanced firmware authoring resource (see ultimaterobolab.com for more information).
- www.philohome.com. Many inspirational projects and detailed resources.
- www.teamhassenplug.org. A clearinghouse of resources and ideas!

- www.extremenxt.com/lego.htm. Resources related to Michael Gasperi's work. Check out the section on homebrew sensors. NXT and RCX sensors designed by Michael and many others are represented here. Look here if you can need a custom solution for an investigation or engineering project!

Where to purchase LEGO MINDSTORMS kits, MINDSTORMS NXT Software, and ROBLAB Software:

- www.legoeducation.com. The LEGO group's United States online store.

Third party sensors:

- www.vernier.com/sensors
- www.hitechnic.com
- www.mindsensors.com
- www.dcpmicro.com

LabVIEW™ resources:

- www.ni.com/labview. National Instruments LabVIEW website.
- zone.ni.com/devzone/cda/tut/p/id/4435. Information and resources for the NXT Toolkit, a LabVIEW programming environment for the NXT.
- legoengineering.com/component/content/article/105. More information and resources related to the NXT Toolkit.
- www.extremenxt.com/lego.htm. Look here for more information about Michael Gasperi's book on LabVIEW for the NXT.

LEGO CAD:

- www.ldraw.org. Extensive LEGO CAD site. Offers resources for multiple computer platforms.
- ldd.lego.com. Official LEGO CAD program.

Books:

From College House Enterprises, LLC:

- Physics by Design, 2nd Edition: ROBOLAB Activities for the NXT and RCX
  Physics by Design 3rd Edition: with NXT MINDSTORMS
  By Barbara Bratzel. An extensive source of upper middle and lower secondary school physics activities.
- Engineering With LEGO Bricks and ROBOLAB by Eric Wang.
  The second edition focuses on the RCX and Robolab. The third edition focuses on the NXT and ROBOLAB An extensive reference book for building LEGO robot models and programming them with Robolab.

- LEGO NXT-G For ROBOLAB users by Eric Wang. This is an excellent migration guide for those switching from the ROBOLAB programming environment to the MINDSTORMS NXT Software (NXT-G) environment.

From Vernier Software & Technology:

- STEM With Vernier and LEGO MINDSTORMS NXT. Activities for using the NXT and Vernier sensors in science, technology, engineering, and math (STEM) lessons.

NOTE: These are only a few of many books available on the topics addressed in this book. Visits the websites noted here to find many more excellent titles.

Chapter 3

Getting Started: Part II. **Software.**

3.1 Introduction

In this chapter, we provide basic tutorial information to get started with four different software options for programming and data logging in robotics projects. As noted in Chapter 2, these are not the only options for programming NXT or RCX hardware; these are the programs we have used most with our students. They provide excellent entry level as well as advanced options for teachers and students.

3.2 LEGO® MINDSTORMS® Education NXT Software 2.0

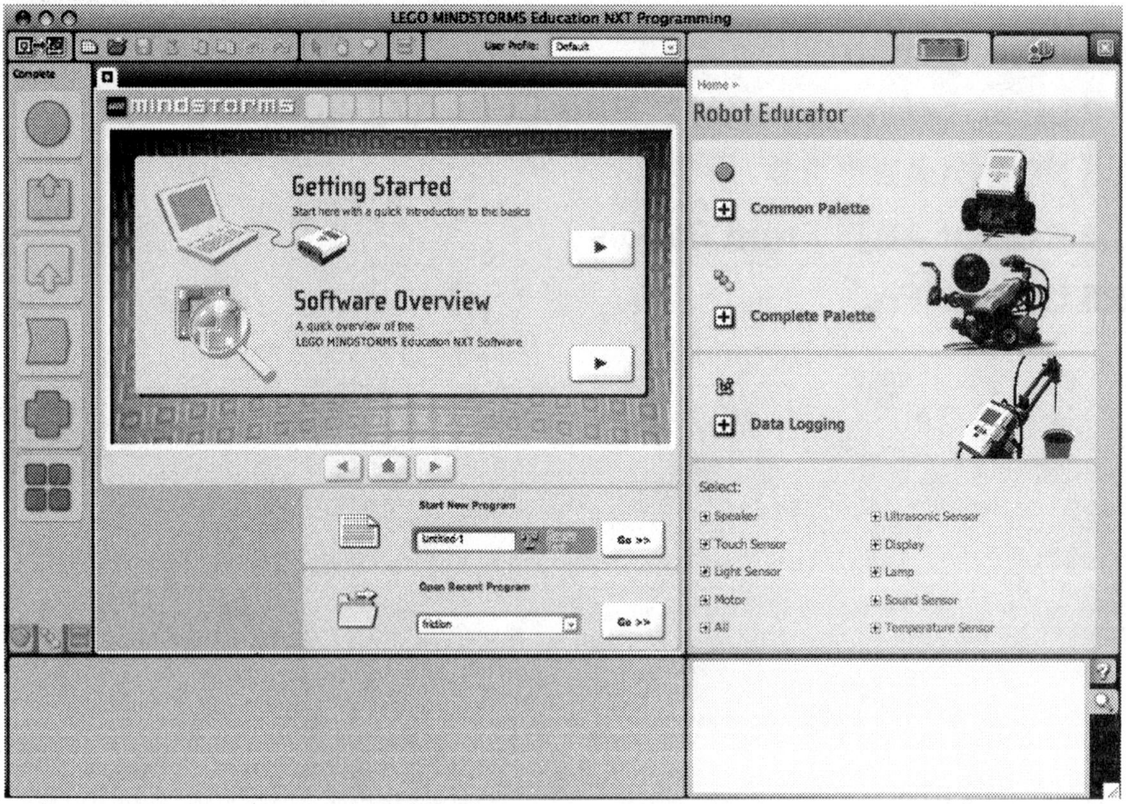

Figure 3.2.1 Opening screen for LEGO MINDSTORMS Education NXT Software 2.0.

What follows is a brief introduction to the LEGO MINDSTORMS Education NXT Software 2.0. MINDSTORMS software is a graphics based program built on the industry standard program LabVIEW™ from National Instruments (www.ni.com). What sets this program apart from other software available to program the NXT are the embedded objects called "blocks". When a block is chosen from the menu (called a palette – Figure 3.2.2) and active on the work page, a configuration window appears at the bottom of the page with settings that can be altered. The combination of easy to interpret graphical blocks and a configuration window allows students and teachers to start programming with little to no prior experience.

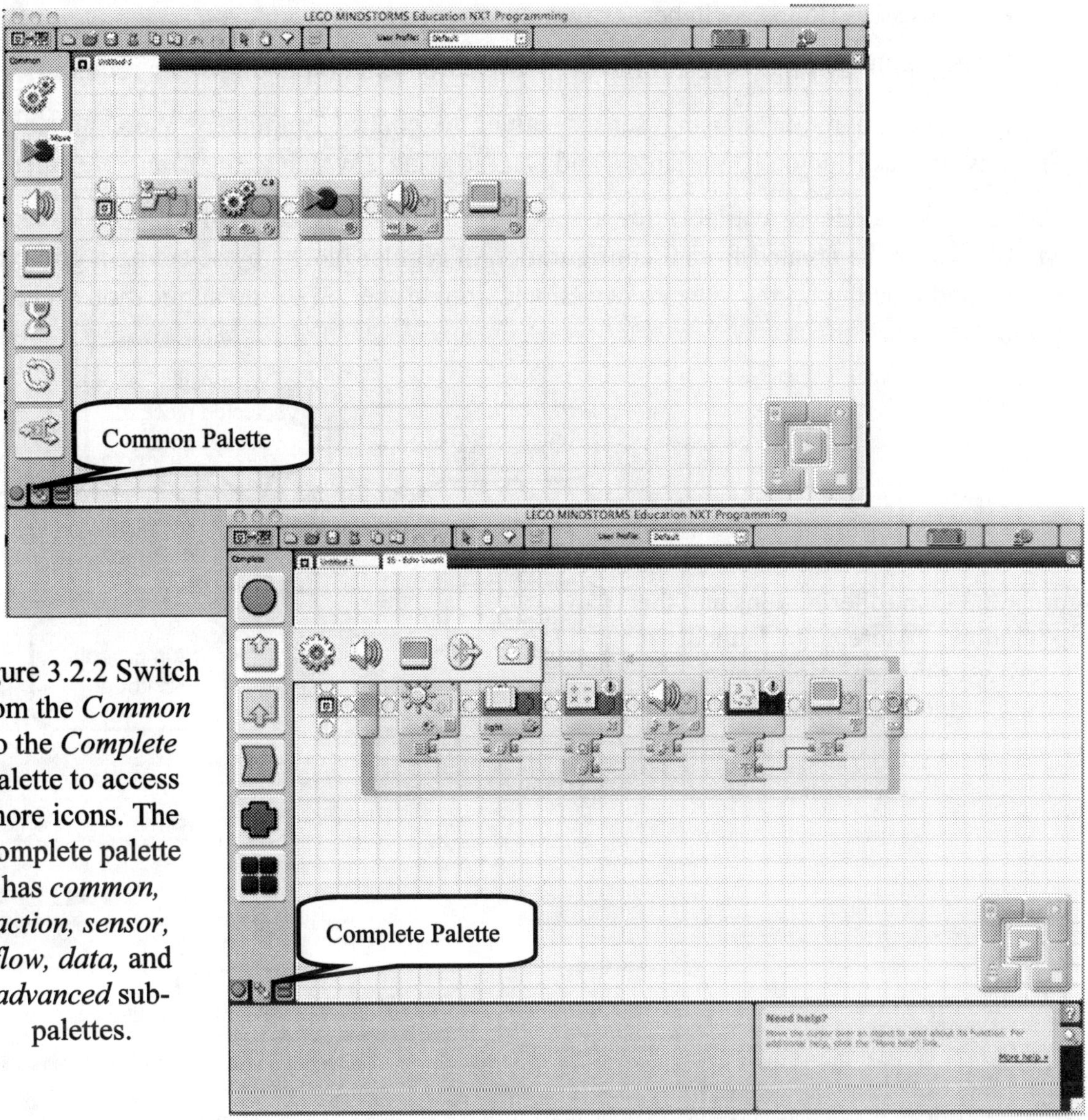

Figure 3.2.2 Switch from the *Common* to the *Complete* palette to access more icons. The complete palette has *common, action, sensor, flow, data,* and *advanced* sub-palettes.

Most of the programs you write will require multiple blocks. Blocks are linked along a data wire that looks like an extendable LEGO beam, with positions made available as blocks are moved. This makes programs very easy to write and requires few objects because of the extensive configurations built into each block. Fairly complex programs can be written with only a few icons.

Program blocks are organized into three palettes -- the *Common* palette, *Complete* palette, and *Custom* palette. The common and complete palettes are suited for beginner and intermediate levels of programming. The Common palette includes the basic blocks used in robotics projects:

- Outputs (motor, sound and display)
- Inputs (wait for sensor input or time)

- Processors (loops, switches, or forks)
- Record/Play (allows recording of motor actions and replaying when required)

If your program requires using sensor values to control outputs or more advanced features such as Bluetooth, Variables, and Data logging[1], switch to the *Complete* palette.

The *Custom* palette is available for a special group of programs called "My Blocks". My blocks can be created from a configuration of regular blocks by selecting a group of blocks and clicking on the "Create My Block" tool at the top of the application window (Figure 3.2.4).

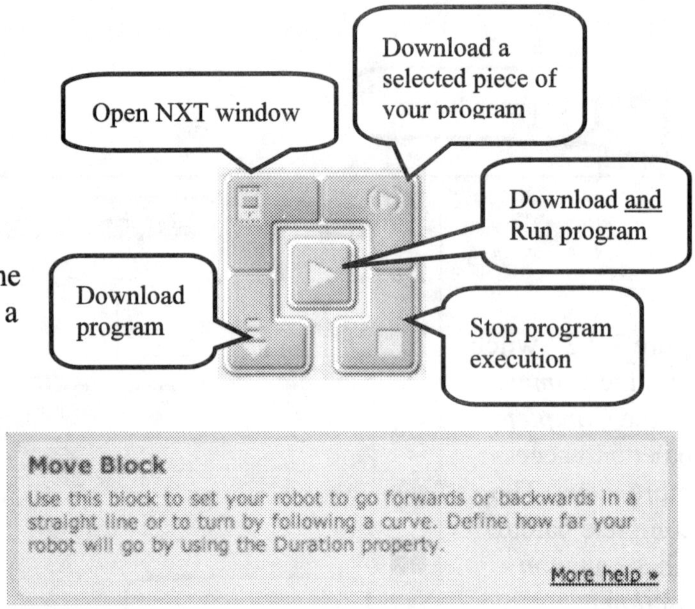

Figure 3.2.3 Controller buttons and the help window. "More help" launches a browser to view application documentation.

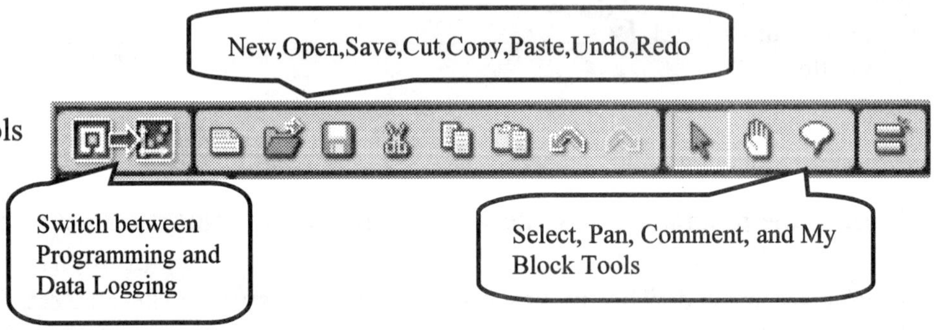

Figure 3.2.4. Tools available at the top of the application window.

[1] Available only for the 2.0 version of the MINDSTORMS software.

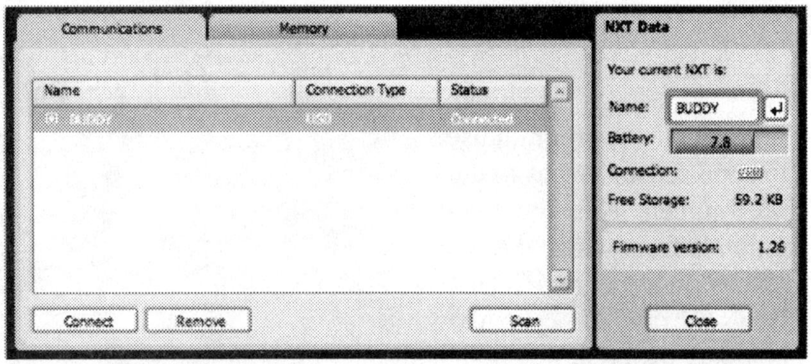

Figure 3.2.5 The NXT window button on the controller is used to manage your device.

Using Sensors

In the *Complete* palette is a sub-palette called *Sensor* and a list of supported sensors are shown when it is accessed. If you drag a sensor out of a palette and into the programming area, you will see a display area for the sensor in the lower left hand corner of the screen. You can see the current value of the sensor in this display area if the NXT is on and connected (Figure 3.2.6). This can be very useful for checking sensor values or even monitoring an event using the computer as a digital display for sensor connected to the NXT.

To use the NXT screen as a display for sensor readings, you can add the "display" icon from the *Action* sub-palette. An example program is shown in Figure 3.2.7. One thing to notice in the example program from Figure 3.2.7 is the flow of data. The value of a sensor can be sent to different parts of a program by wiring sensor values to other icons in your program.

When used in conjunction with icons from the *Data* palette, sensor values can be wired into logic statements, math functions, value comparisons, range determinations, and stored in variables. In the *Flow* sub-palette, sensors are built into wait statements, while loops, and switches (if/then statements).

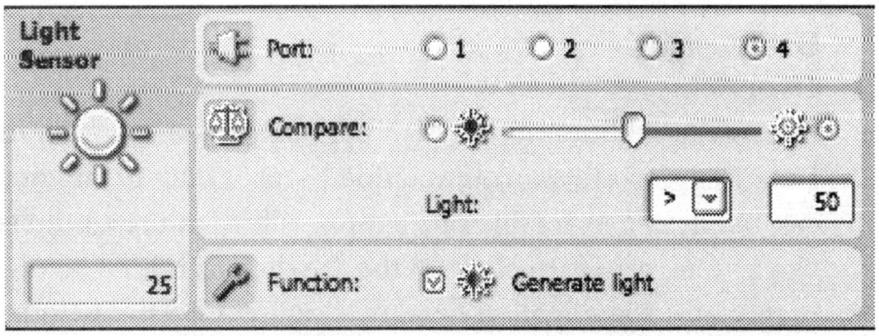

Figure 3.2.6. Sensor display. Note the "live" value of the light sensor (25) below the sensor icon.

Figure 3.2.8 shows an example program in which the value from the light sensor is multiplied by a constant and then wired into the speaker icon. The result is a tone that is proportional to the value of the light sensor. Converting sensor values to auditory tones gives students and teachers a very unique way to observe phenomena.

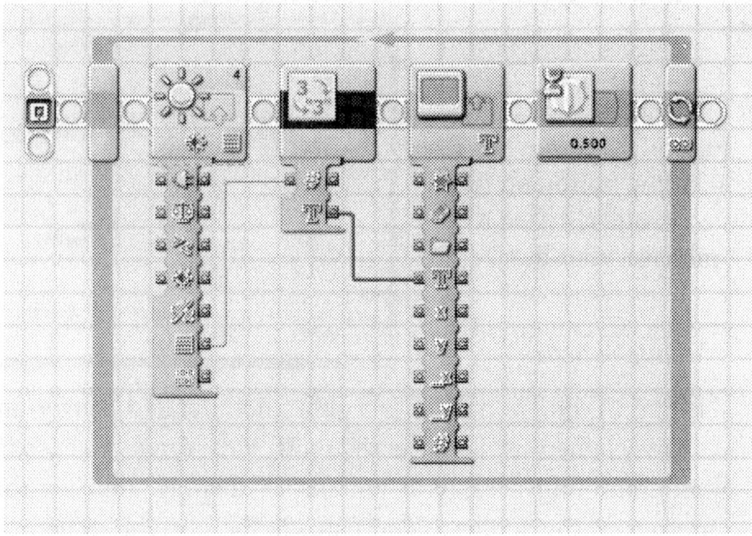

Figure 3.2.7. A program used to send sensor values to the NXT screen. Note the many terminals from which data can "flow" in a program. This allows for program control on many different levels.

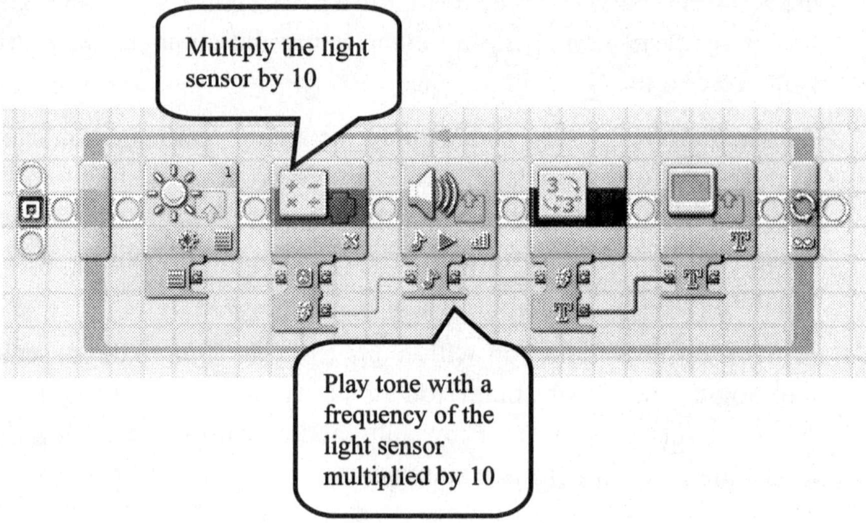

Figure 3.2.8. In this program, light values are converted into sounds.

Data logging
Overview

One major addition to the MINDSTORMS software in the 2.0 version is Data Logging. Data Logging allows you to upload sensor data from your NXT to the computer where it can be viewed as a table or a graph. This feature of the software can be accessed in two ways. One way is to click on the "Switch to NXT Data Logging" button in the upper left corner of the application window (Figure 3.2.9). The other is to use a data logging block in a program. The data logging block is found in the "Advanced" category of the *Complete* palette. In this section, we will introduce both data logging methods to you.

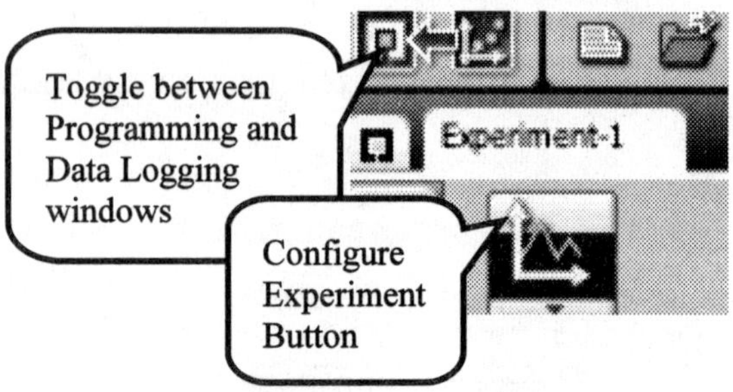

Figure 3.2.9. Once you switch to the Data Logging window, use the Configure Experiment button to define data collection parameters.

<u>Data Logging Window</u>

After you click on the "Switch to NXT Data Logging" button, you will see a window titled, "LEGO MINDSTORMS Education NXT Data Logging". In this window, you can either start a new project or open a saved project. If click on the "Go" button to start a new experiment, you will see an experiment configuration window (Figure 3.2.10). Use this window to tell the system what sensors you will use, where they are located, how fast to collect data and the experiment duration. Once you click OK, you will see a blank graph and you are ready to go. Connect your sensors, turn on your NXT, and start your experiment by pressing the "play" button on the data logging controller (right and side of the screen). If you need to change parameters from the experiment configuration window during an experiment, you can access this window again by clicking on the "Configure Experiment" button on the top-left side of your graph (Figure 3.2.9).

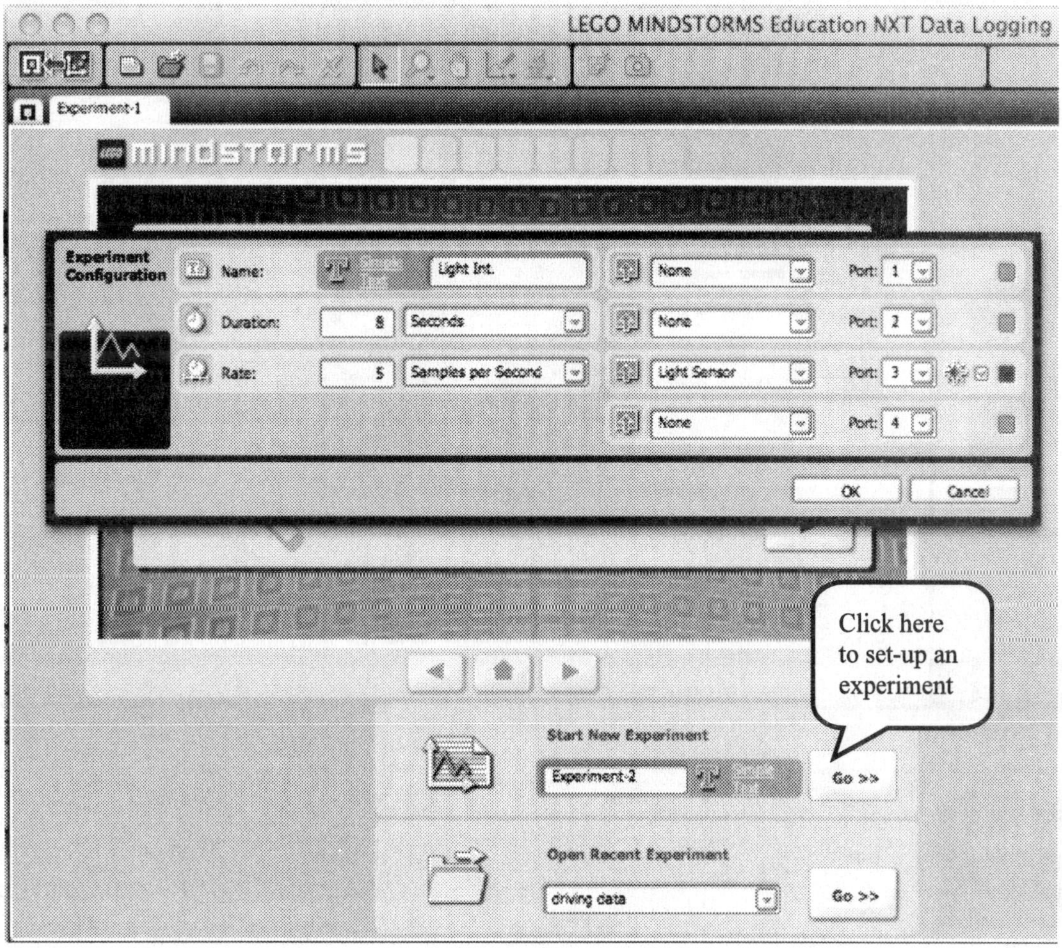

Figure 3.2.10 Data Logging configuration window

At this point, you can interact with the program like most data logging software applications. You can collect data, do a basic analysis (see next section), and save your data. When you save your data, it is saved as "log" file (Figure 3.2.11). The log file is simply a text file. If you need to do additional analysis of the data, you can import the log file into a spreadsheet. If you have difficulty importing a file into a spreadsheet, open it up in a text editor, delete everything that is not your data and resave it as a text file. This will leave you with a "tab delineated" data file that can be readily imported into most spreadsheet applications.

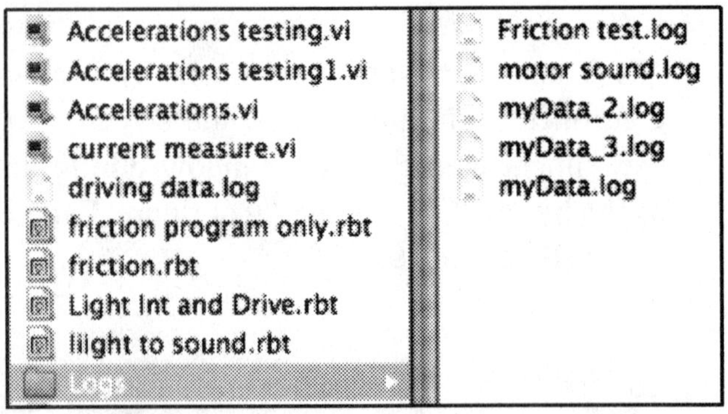

Figure 3.2.11. Data is saved in a .log file format. A .log file can be opened by text editors or even web browsers to see the data.

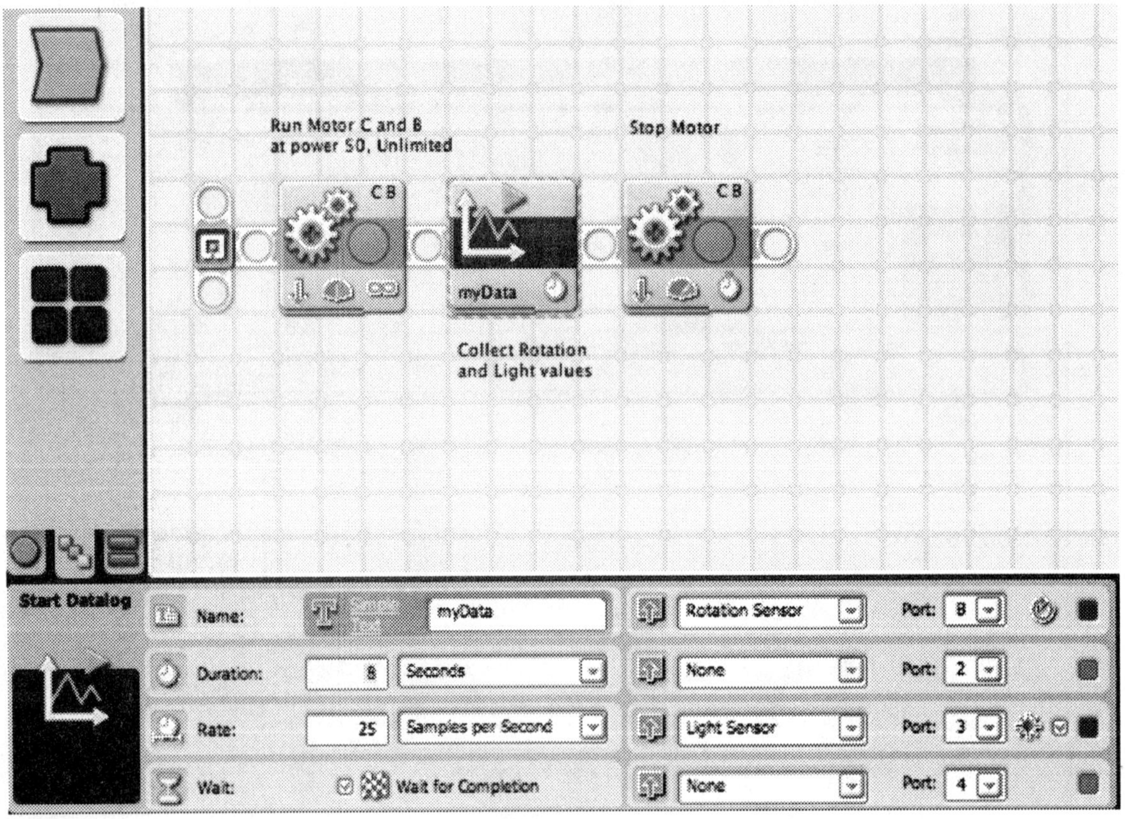

Figure 3.2.12 Sample program to collect data

Example data logging program

One compelling reason to use robotics in science investigations is the ability to write a program to control an experiment. For instance, if you are exploring the relationship between distance from a light source and light intensity, you could follow these steps:

1. Write a program to collect data while your robot[2] drives away from a light source (Figure 3.2.12).
2. Setup your experiment in the Start Data Log block's configuration window. Figure 3.2.12 shows that we are collecting both light intensity and rotation sensor data.
3. Download your program to the NXT, disconnect it, and run your robot away from the light source.
4. Reconnect the robot, switch to the data logging window and click on the Upload From NXT button on the data logging controller. This will bring up the window shown in Figure 3.2.13. Choose your data file and click on "upload" to send it to the computer. You will then see a graph of your data (Figure 3.2.14). At this point, you can choose to use the MINDSTORM software's data analysis tools or move the data to a spreadsheet.

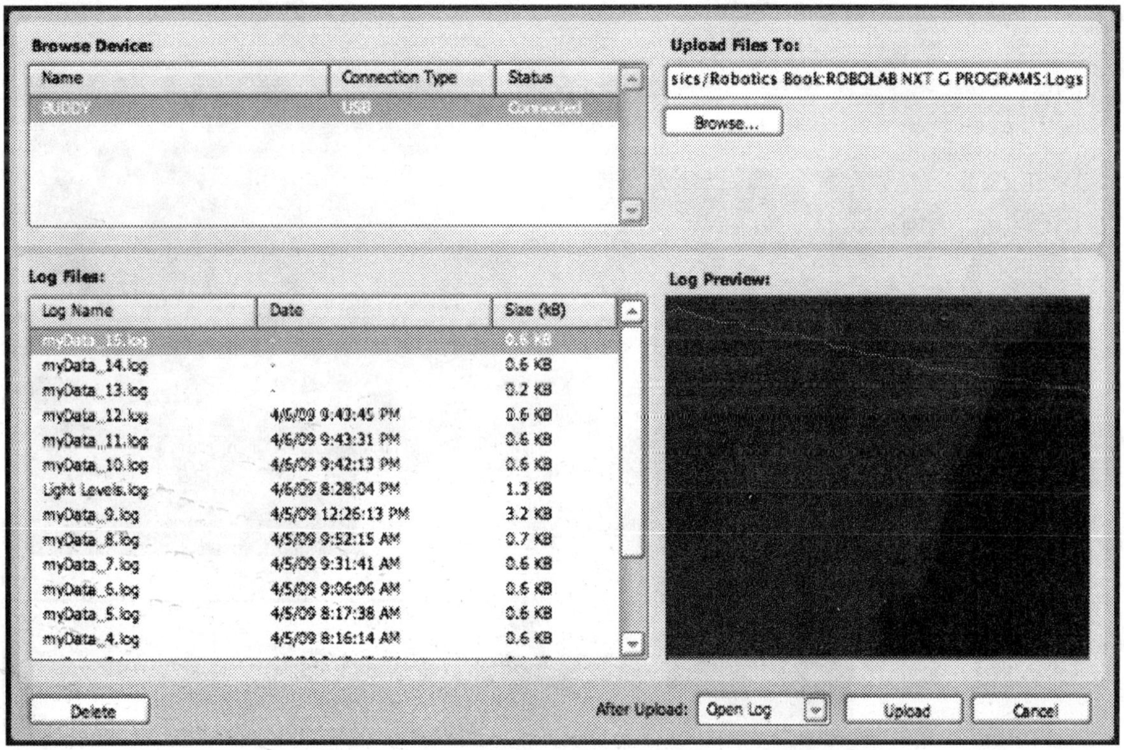

Figure 3.2.13. Data files on the NXT can be managed after clicking the "Upload From NXT" button on the data logging controller.

---

[2] See vehicle design in activity 6.2, "Wave Intensity vs. Distance".

In this example, we have shown that it is straightforward to create an experiment to drive your robot vehicle while collecting light data. Before leaving this example, however, it is worth noting two limitations of using MINDSTORMS in data logging activities. First, note the units used by a sensor. Here, we are using a light sensor and measuring light intensity but our units are a percentage rather than Watts/m^2. This is not ideal if you need actual intensity readings in your experiment. Second, the graph is not of wave intensity vs. distance but rather a dual y-axis graph of percent light vs. time and rotation vs. time (Figure 3.2.14). In the MINDSTORMS software, you cannot change the x-axis.

There are three ways to manage these limitations. First, we ask our students to look for general relationships in the graphs of data rather than explicit values. In other words we ask them to look at "trends" rather than the specific mathematical function. Second, if we need better values, we use better sensors. As we point out in the next section, you can connect third party sensors to your robot and obtain data from a wide array of high quality instruments. Third, if we need to transform the axes of our graphs in ways that are not possible in the MINDSTORMS software, we move the data to spreadsheet program where such transformations are possible. If you wish to explore this example further, see Chapter 6, activity 6.2 -- "Wave Intensity vs. Distance".

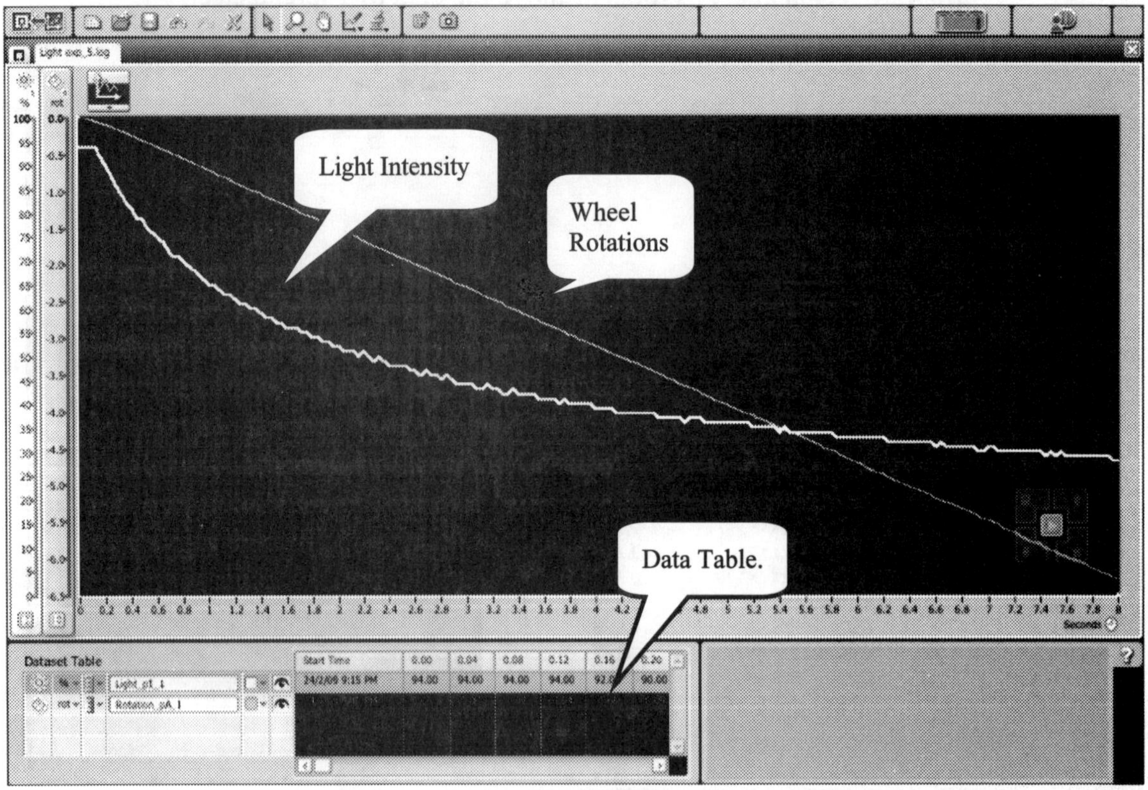

Figure 3.2.14. A graph of light intensity and wheel rotation data in the data logging window of the MINDSTORMS software.

Basics of Data Analysis

Data analysis can be done within the data logging window (Figure 3.2.15). Capabilities include:

- Multiple data sets plotted on one graph
- Change of scale for all y axis variables
- Prediction tools (linear function and free drawing)
- Min, max, and mean of a selected region
- Linear fit of selected region

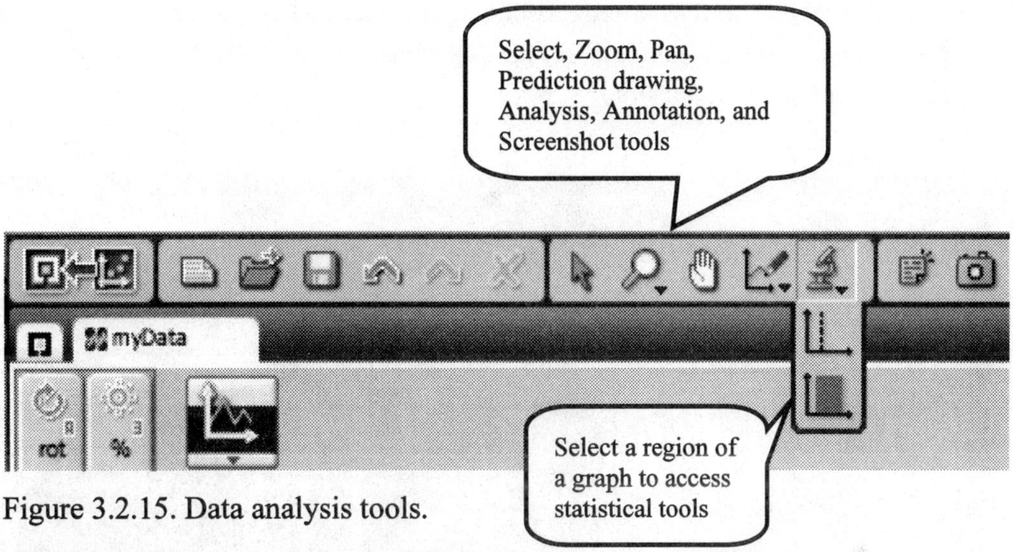

Figure 3.2.15. Data analysis tools.

Using Non Lego sensors with Mindstorms

Programming blocks are available from suppliers of other sensors that connect to the NXT:

1. Vernier Sensors using the Vernier Sensor Adapter:

   See http://www.vernier.com/nxt/

2. HiTechnic Sensors:

   See http://www.hitechnic.com/

Consult the instructions provided with these blocks for further information about their use in programming and data logging projects.

To go beyond the information presented here on the MINDSTORMS software, visit www.mindstorms.lego.com and other resources found at the end of Chapter 2.

3.3 ROBOLAB™ 2.9

Originally developed in 1998, ROBOLAB was designed as a programming and data logging environment for the RCX. It was the result of a partnership between Tufts University's Center for Engineering Education and Outreach (CEEO), National Instruments, and LEGO Education. In 2006, ROBOLAB was updated to work with both the RCX and the NXT. In 2006, MINDSTORMS NXT Education Software was also introduced with the idea of replacing ROBOLAB at the primary and middle school levels. As we presented in our previous section, however, MINDSTORMS software is also an excellent tool for high school physics instruction. What it lacks, though, are advanced data analysis and customization features needed by upper high school and college physics instructors. You will see in this chapter how ROBOLAB is designed for both beginners and advanced users.

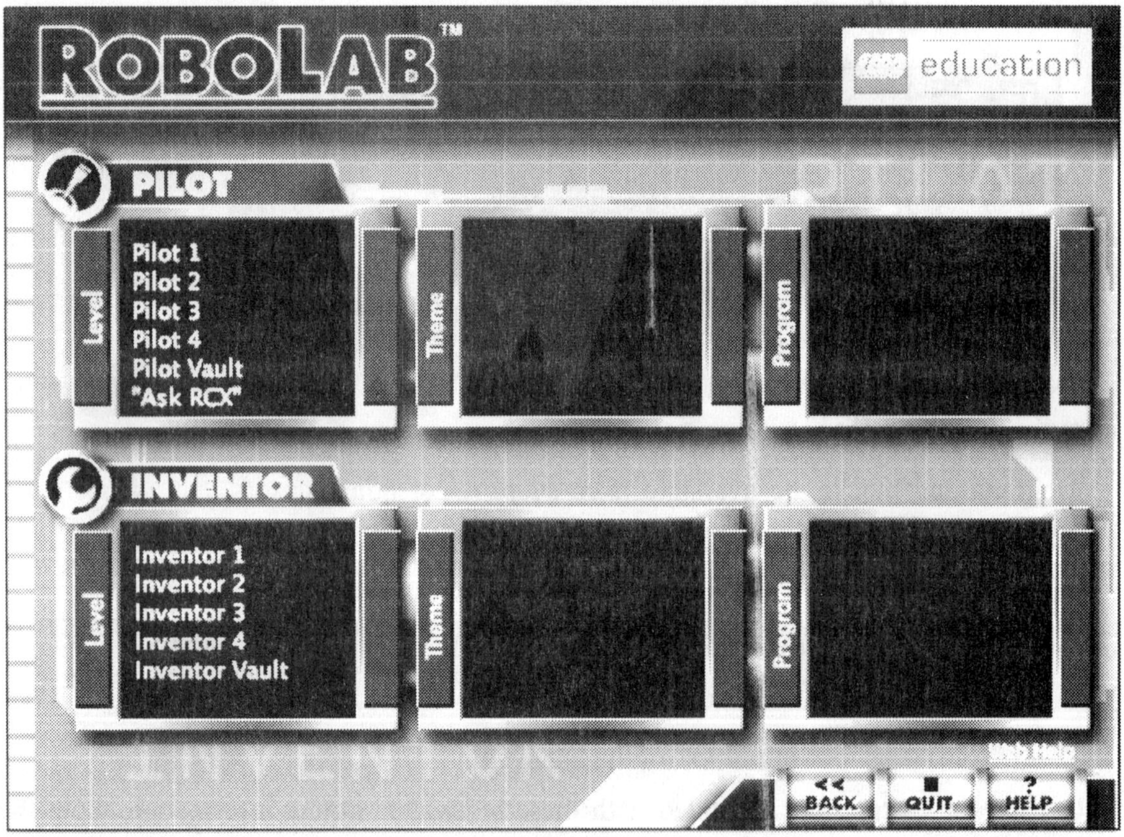

Figure 3.3.1 ROBOLAB's Programmer environment is organized into 4 "Pilot" levels and 4 "Inventor" levels. If you need to write programs quickly for motors and a few sensors, use the Pilot levels. These levels are great for many of the activities in the book. If you need to write custom programs with specific goals that cannot be reached in Pilot mode, use the Inventor levels.

We provide a general overview of ROBOLAB in section 3.3, documenting how to complete programming and data logging tasks. In section 3.4, we provide an overview of how to access advanced ROBOLAB features by instructing the reader to create custom "front panels". Like the MINDSTORMS software, ROBOLAB is also based on LabVIEW. One key difference between these software packages is that ROBOLAB allows the user to access LabVIEW tools such as custom front panels. If a high school version of LabVIEW is of interest to you, the Tufts CEEO and National Instruments are currently developing a new product called LabVIEW for High School. This product will build upon the success of ROBOLAB and give secondary teachers and students more access to the powerful LabVIEW package. We are getting ahead of ourselves here but felt it worth noting the relationship between ROBOLAB and the future LabVIEW for High School. The MINDSTORMS software was designed to succeed ROBOLAB in the primary and middle school market and LabVIEW for High School will succeed ROBOLAB in the high school market. ROBOLAB, however, will still live on. Once LabVIEW for High School is released, ROBOLAB will be made available as an optional download at http://www.ceeo.tufts.edu/robolab. For the most current information on this topic, visit our website, http://www.physicswithrobotics.com

As stated earlier, this section on ROBOLAB provides a brief overview of the ROBOLAB programming language and its data logging capabilities. If you need more documentation, it is available when you install the software on your hard drive or from several books that have been written to support teachers using ROBOLAB. See the resources section of Chapter 2 for more information. Finally, visit LEGOengineering.com for additional ROBOLAB learning resources.

When you launch ROBLAB, you will see three choices: Administrator, Programmer, and Investigator. Administrator allows you to manage your RCX or NXT by changing communication ports, updating firmware, changing both software and hardware preferences, and checking your battery level. Like MINDSTORMS software, ROBOLAB provides both programming and data logging environments. These environments are organized into three areas: Pilot, Inventor, and Investigator. Pilot and Inventor are found after pressing the "Programmer" button on the opening screen (Figure 3.3.1).

Programmer: Pilot

Pilot is a very fast and easy to use programming environment. It uses icons that are already placed together as a program. To change your program, click on an icon to choose from a few options (Figure 3.3.1). The default pilot programs make it convenient for the user to program the NXT or RCX with simple behaviors. For instance, the program in Figure 3.3.2 instructs an NXT to turn on two motors and a lamp for two seconds and then change the direction of motors and continue running the motors until the button sensor is pressed. Many physics activities can be done with Pilot level programming.

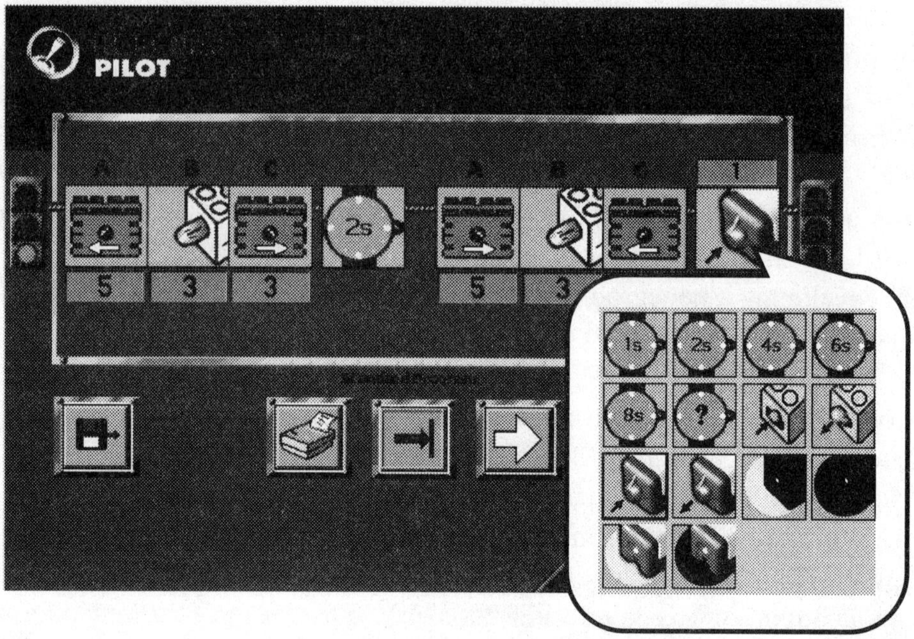

Figure 3.3.2. Pilot levels start with a default program. Programs are changed by clicking on icons and choosing from a menu of choices.

Programmer: Inventor

In this environment, program icons are "wired" together to build a program. It is an extremely flexible and customizable environment. There are four programming levels within Inventor. Default programming structures that can be changed or modified by the user are found in levels 1 and 2. Levels 3 and 4 start with a blank programming window and users must write their entire program from icons in the "Functions palette" (Figure 3.3.3). This palette provides a selection of programming elements such as sensor inputs, motor and lamp outputs, program variables, loops and other programming structures.

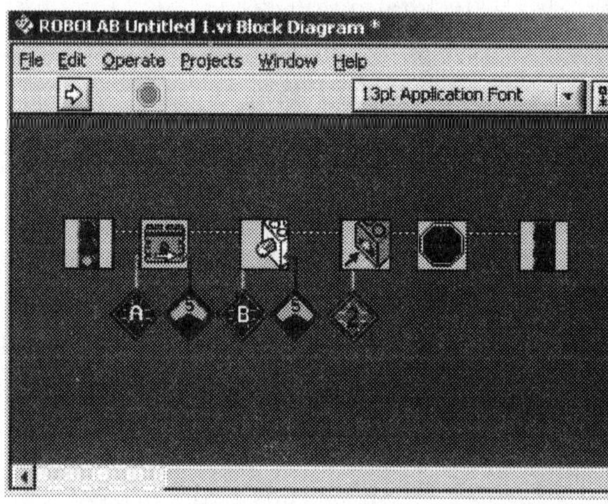

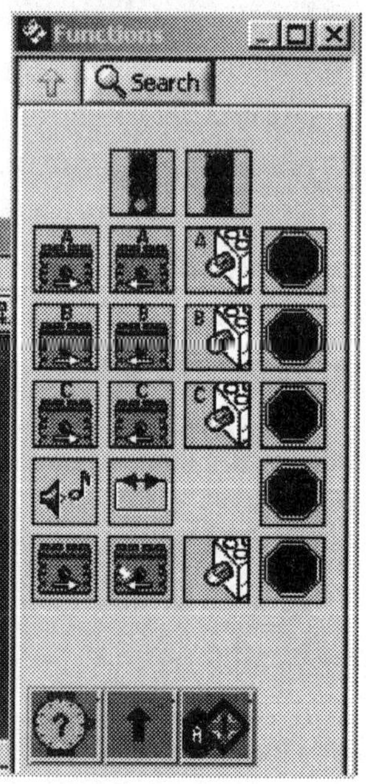

Figure 3.3.3. Inventor levels require users to wire icons together from a palette of choices. Shown here is Inventor Level 2.

The main difference between Inventor levels is the number of icons available in a palette. Figure 3.3.4 shows the palette for Inventor level 4. This level offers you many options but requires more time to learn how to take advantage of the available programming power. If you anticipate that you and your students will be creating custom programs and are interested in pushing your hardware and software to the limits then taking the time to learn how to program in Inventor level 4 is well worth the effort. One guide to help you choose which level to use is the consideration of programming speed and complexity. The higher Inventor and Pilot levels require more time but you will have access to more options. In our experience with ROBOLAB, we tend to use the Pilot levels or Inventor 4. By starting with Inventor 4, you know that you have access to all the features that you would ever need. For instance, access to 3rd party sensors is only available in Inventor 4.

3rd Party Sensors

If you wish to use sensors from 3rd party vendors, you will be required to write your programs in Inventor level 4. If your system is updated to version 2.9.4c, you will have icons in your "User Libraries" folder (see lower call-out in Figure 3.3.5) for sensors from Vernier, HiTechnic, and Mindsensors[3].

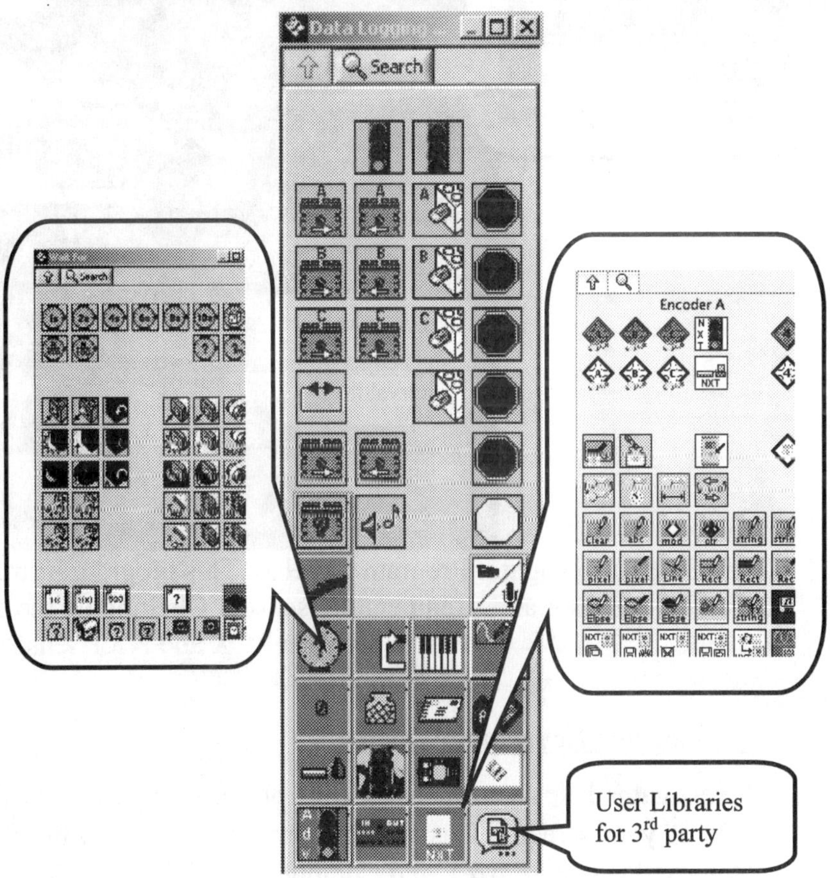

Figure 3.3.4. Inventor 4 icon palette. The bottom 16 buttons in the palette reveal more options. Shown here are icons for timers, over 10 RCX and NXT sensors, image processing, NXT screen commands, motor commands and other advanced features.

[3] Visit LEGOengineering.com for ROBOLAB updates.

Data Logging Overview: Investigator

The Investigator programming mode includes features from both Pilot and Inventor levels along with the added feature of data logging and advanced data analysis. There are options for 5 programming levels within the Investigator. Levels 1-3 provide the user with a default data logging program. For example, the level 2 program shown in Figure 3.3.5 tells the RCX to collect light sensor data and temperature data at 1 second intervals while Motor A is running. The program is stopped after 10 samples are collected. This program could be used on a single motor robot when you need to collect data as the vehicle is driving. Changes are made to programs by clicking on icons and picking new ones (Figure 3.3.5). Level 4 and 5 programs require the user to write their own programs, similar to Inventor programming (Figure 3.3.7). Collecting data from 3rd party sensors is only possible in Investigator Level 5.

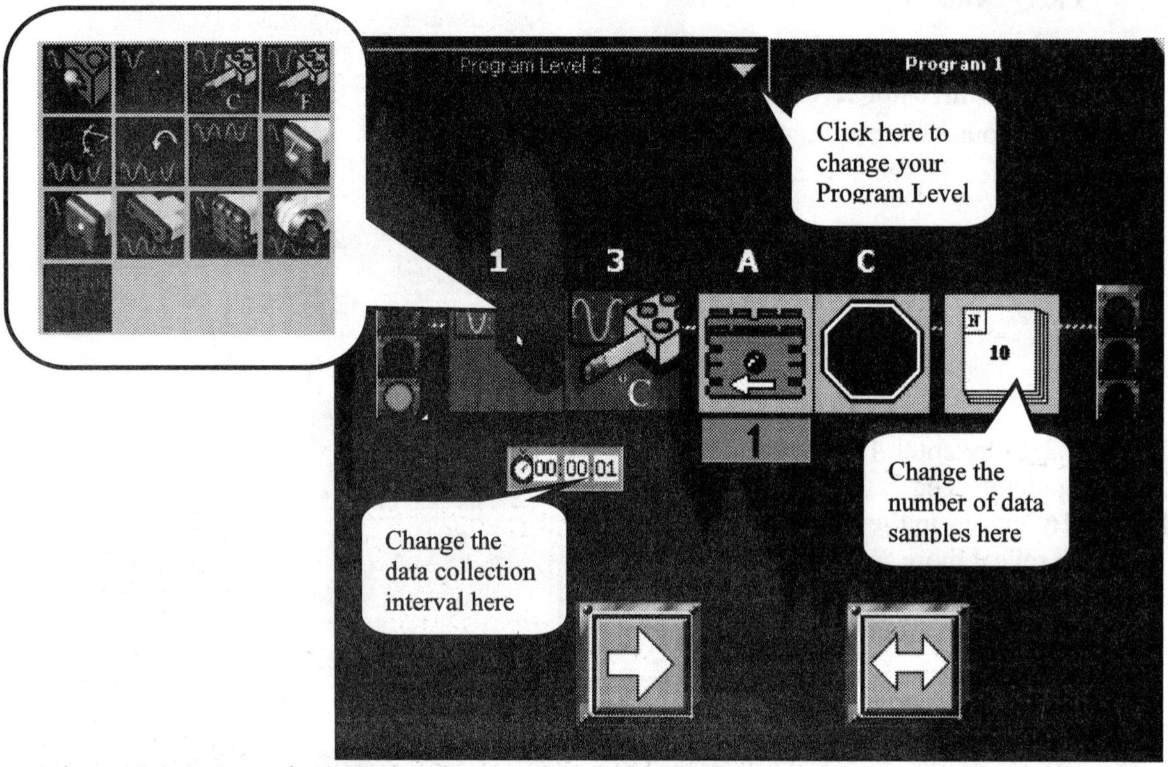

Figure 3.3.5. Investigator Program Level 2. This program instructs an RCX to collect data from a light sensor and a temperature sensor. Clicking on the sensor icons allows you to change between RCX and NXT sensors.

<u>Data logging Navigation Window</u>

Investigator helps you manage your projects through a navigation window on the left side of your screen. This window provides access to the Program Area, Upload Area, View and Compare Area, and Compute Tools (Figure 3.3.6).

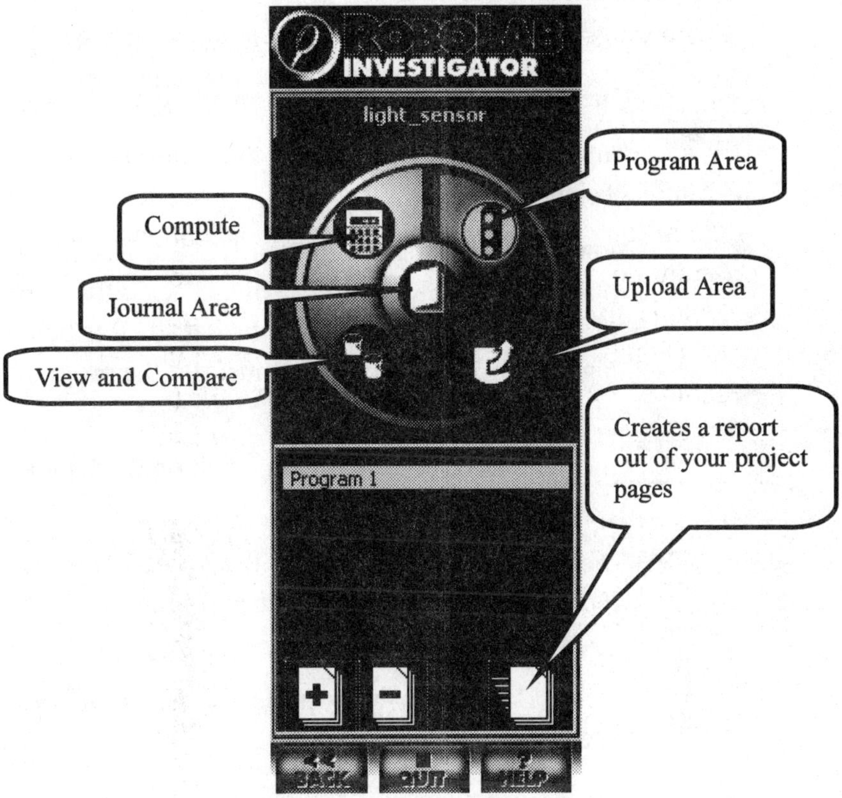

Figure 3.3.6. The navigation window is used to manage data logging projects.

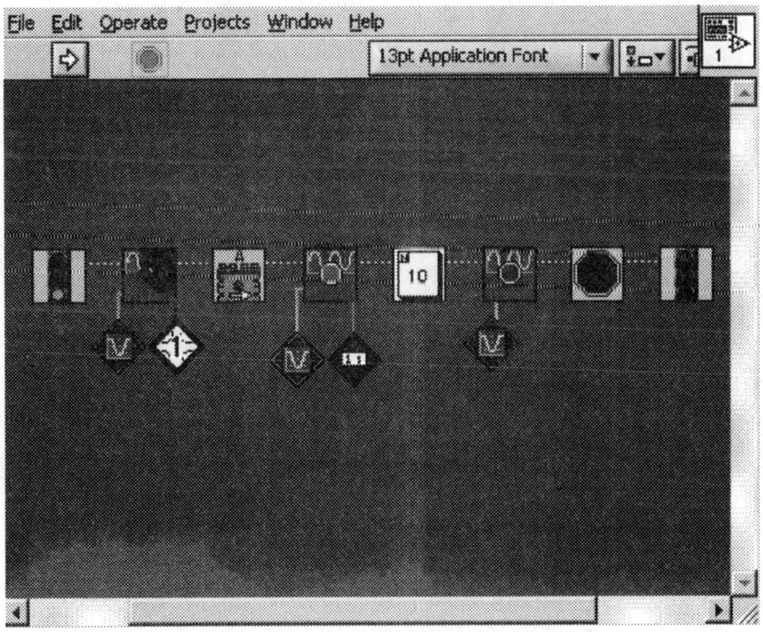

Figure 3.3.7 Investigator 4 and 5 programs benefit from the expanded palette of programming icons found in the Inventor programming levels. This allows for great control over how data is collected and managed.

Example data logging program

Follow these steps to collect and graph data with your RCX or NXT:

1. Open up a new Investigator project. Choose Program Level 1.
2. Connect desired sensors. Set the desired data collection interval and number of samples.
3. Download the program to your RCX or NXT by either clicking on the single white arrow or the double arrow. The double arrow will run the program in "direct mode", which graphs the data as it is being collected. If you use the single download arrow, you must run the program on the RCX or NXT to collect data.
4. Once you have collected data, click on the Upload Area in the Navigation Window.
5. Click on the white arrow in the upload arrow to transfer the data from the RCX or NXT to your computer. Figure 3.3.8 shows the upload area with an example graph.
6. To add Data Sets to your project, click on the "+" in the upload area and repeat steps 1-5.

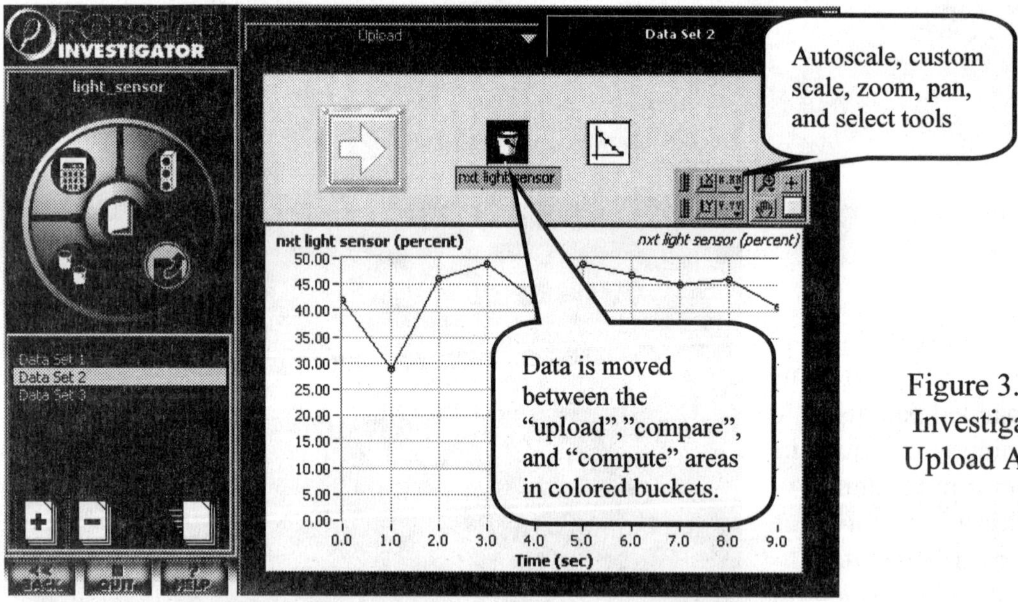

Figure 3.3.8. Investigator Upload Area.

Basics of Data Analysis

View and Compare

The View and Compare area allows you to view uploaded data sets and obtain statistical information. Data can be viewed in different representations such as line, bar, and table form. Several statistical tools are available including measurements for minima, maxima, means, standard deviation, slopes, and areas (Figure 3.3.9).

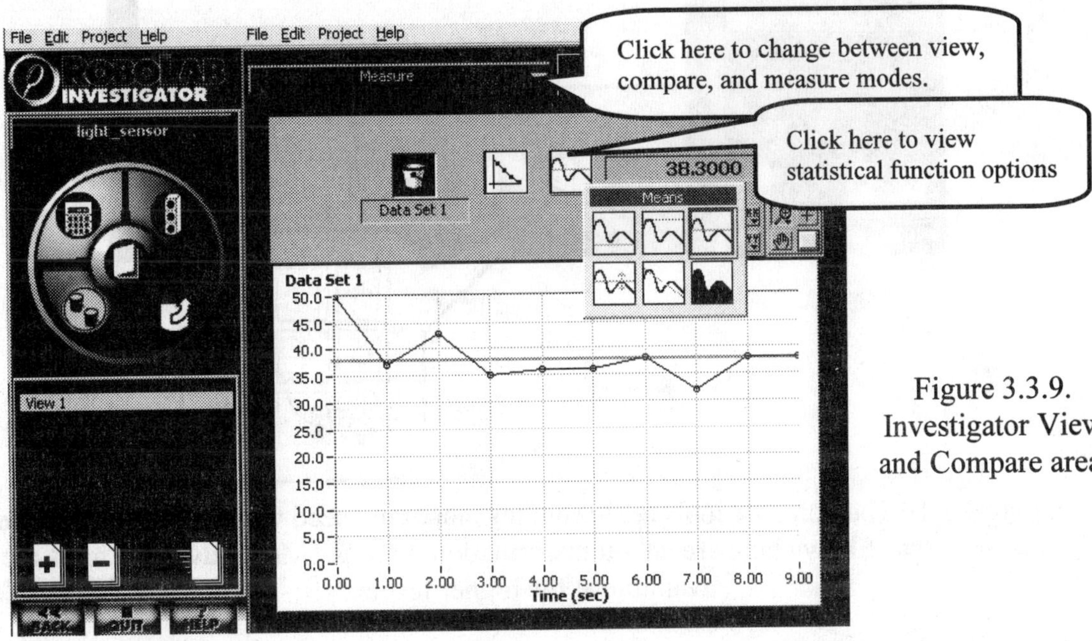

Figure 3.3.9. Investigator View and Compare area.

Compute Tools

The compute tools area provides additional options for data analysis. This area is used when calculus and other advanced functions are required. This option has five levels. Compute tools levels 1-3 include analysis options for data transformations (level 1), changing the x-axis from time to another data set (level 2), statistics to more complex operations such as taking derivatives and integrals (level 3), and writing your own analysis program using LabVIEW tools. Figure 3.3.10 shows compute tools level 1. Figure 3.3.11 shows compute tools level 5.

If you are interested in utilizing ROBLAB's advanced data analysis features, look at the documentation documents in the ROBOLAB 2.9 application folder on your computer. The files are named:

- Using QuickStart_NXT.pdf
- Rl-ReferenceGuide.pdf
- ROBOLAB.pdf

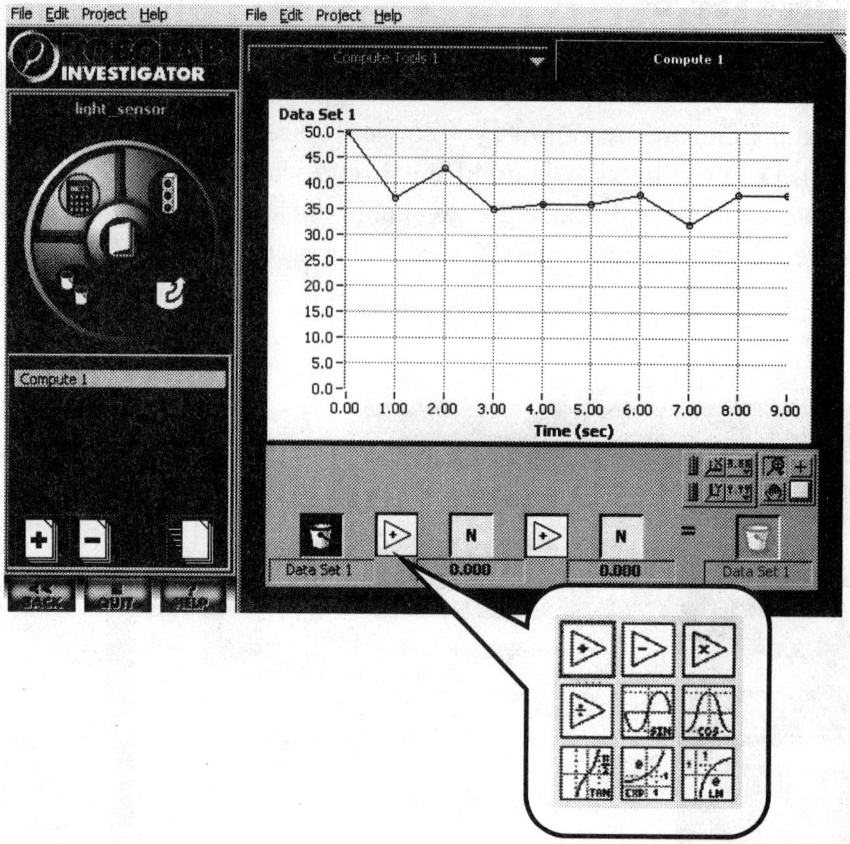

Figure 3.3.10 The compute tools area provides more advanced tools than the view and compare area. Shown here are data transformation options. More advanced tools are available at the higher levels.

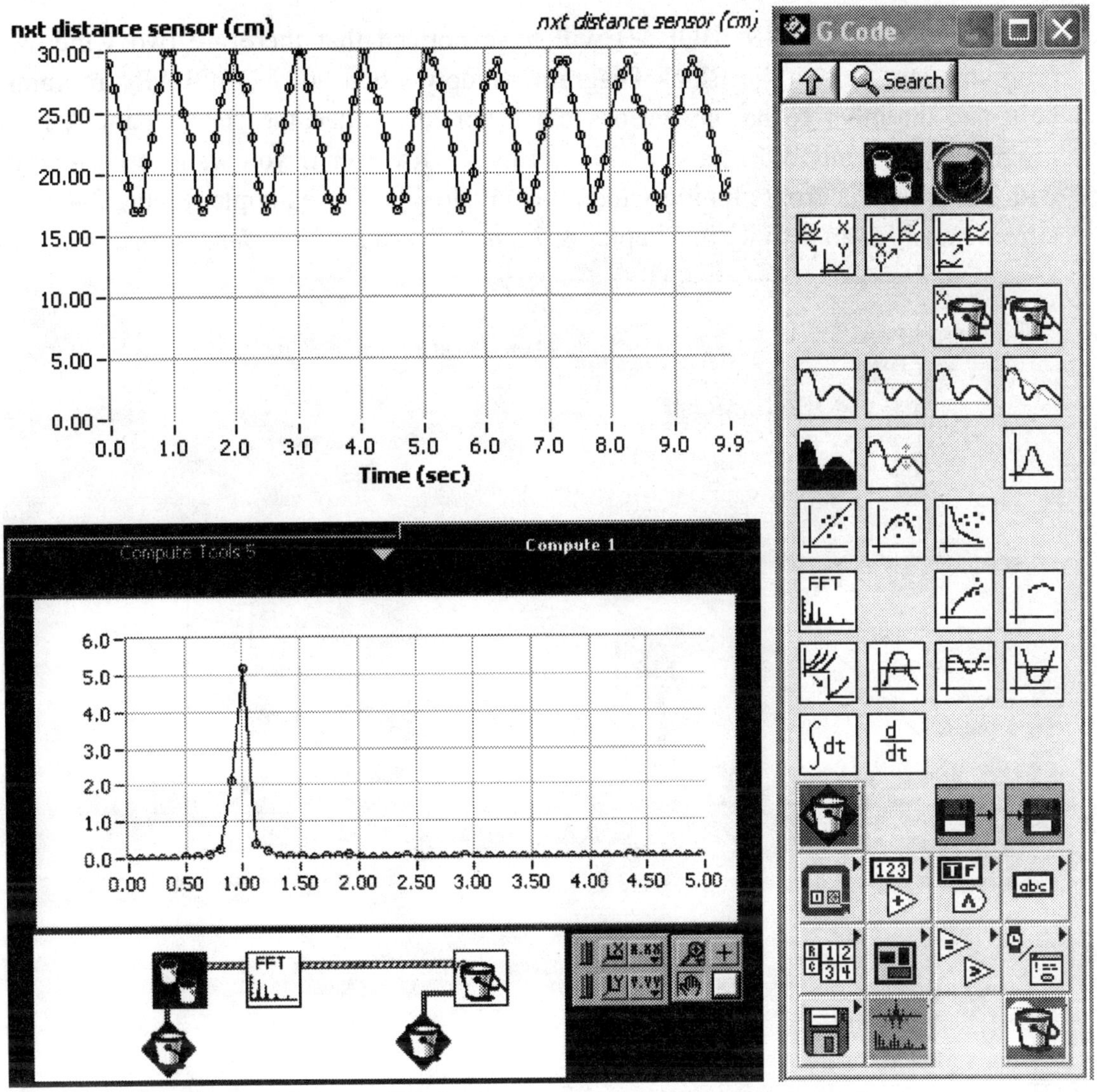

Figure 3.3.11. Compute Tools level 5 allows users to analyze data using advanced LabVIEW tools. In the above example, original data (top) has been analyzed using a Fast Fourier tool (FFT) to find the dominant frequency. Note: in the resulting power spectrum above, the x-axis label should be Frequency (Hz). In this case, the dominant frequency is approximately 1 Hz.

3.4 ROBOLAB™ Front Panels

Those who have used ROBOLAB will have noticed that there are two windows for programming. One is the Block Diagram window where all ROBOLAB programs are built and the other is that mysterious black window called the "Front Panel". In most programs that window is blank and is not used. However if you want to be interactive with the user, this programming window is invaluable. For example, Figure 3.4.1 shows a front panel for an activity in Chapter 4: *Testing Speed and Acceleration of Drag Cars.*

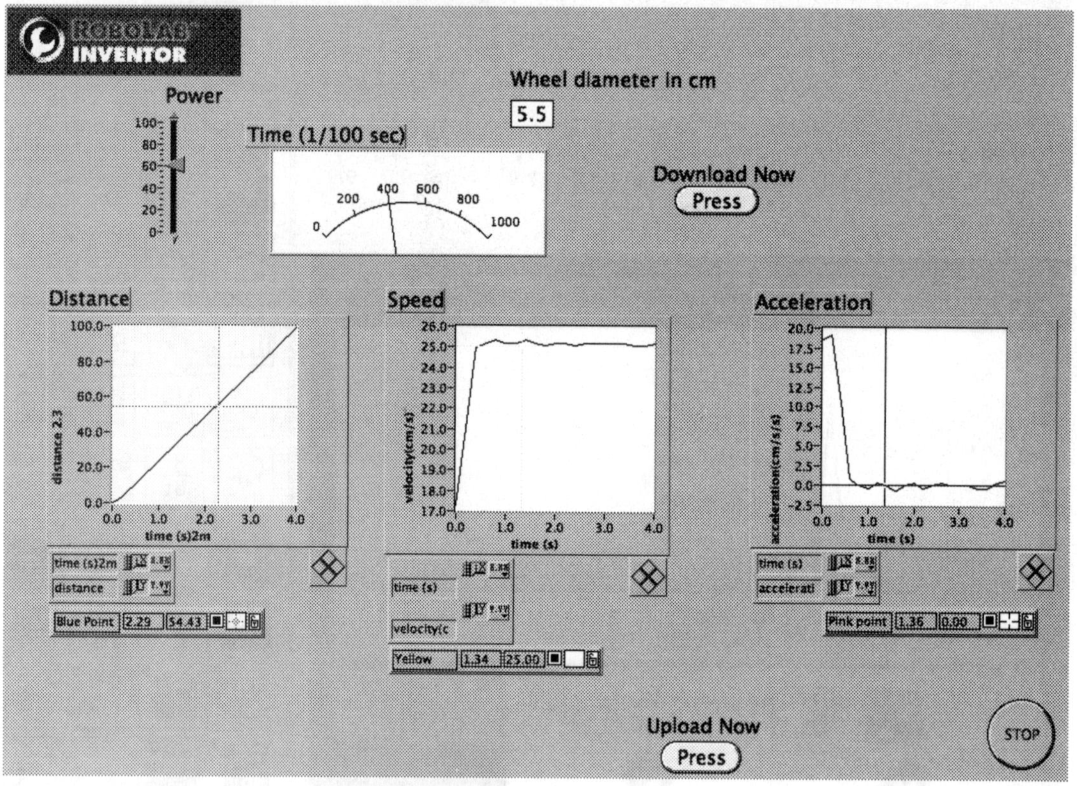

Figure 3.4.1. ROBOLAB Front panel for the Chapter 4 activity: *Testing Speed and Acceleration of Drag Cars.* With custom front panels, all the programming and data logging controls can be placed in one user interface. You can download either RCX or NXT versions of this front panel at our website.

Creating your own front panels is a topic for advanced users. What we will do here is give you a quick overview of how we created the front panel in Figure 3.4.1. If you would like more information about creating custom front panels, Eric Wang's Engineering With LEGO Bricks and ROBOLAB is an excellent starting point. If you do not have ROBOLAB, take a look at Michael Gasperi's book, "LabVIEW for LEGO MINDSTORMS NXT". LEGOEngineering.com also has resources available – search for "LabVIEW" on the website's opening page. Finally, when the previously mentioned "LabVIEW for High Schools" product is finished, more resources will be available.

While creating front panels is advance, using front panels is not! Using front panels with students requires very little programming knowledge. In fact, very little understanding of the operation of the software is required. Simply load and run the front panel. This means you can maximize the time spent data logging, observing and analyzing information in your science project. To help you get started with front panels, we will supply you with the panels linked to some of our activities. For more information, visit our supporting website: www.physicswithrobotics.com.

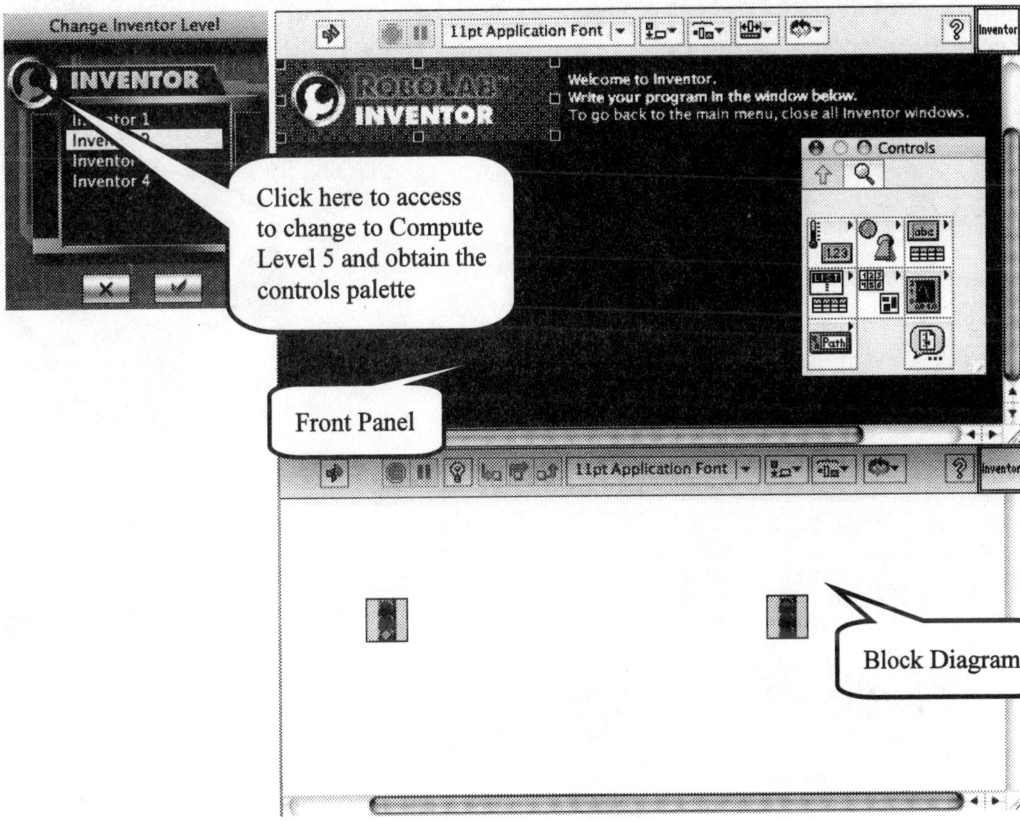

Figure 3.4.2. Every ROBOLAB Inventor program has an unused "front panel" in the top window. To turn this into a functional front panel with an assortment of controls (inputs) and indicators (outputs), choose Project → Change Inventor Level from the application menu bar. Click on the wrench and choose Compute Level 5. Now, when you choose your front panel, you can obtain the controls palette by choosing Window → Show Controls Palette.

Basic principles of LabVIEW™ programming

LabVIEW is an industry standard programming language with software, training, and support available at www.ni.com. Professional versions of the software are expensive but student editions are available at reduced prices. A toolkit for MINDSTORMS NXT programming is available from www.ni.com/mindstorms. ROBOLAB, however, has

LabVIEW as a part of its package. In preparing our front panels for selected activities and writing this chapter of the book, we used ROBOLAB[4].

When you create a new Vi, LabVIEW's name for a program[5], you will be given both a front panel and a block diagram. The front panel and block diagram use graphical language programming. The block diagram, you are probably familiar with as a ROBOLAB programmer. This is where the program icons go. The front panel is the interface used by the operator. The programming elements in a front panel are called "controls" (inputs) and "indicators" (meters, text displays, and graphs). These objects are placed on the screen using the controls palette. This is different from the functions palette on the block diagram and only becomes available if you are working on a front panel window (Figure 3.4.2).

Figure 3.4.3. RCX Angle sensor. LEGO Part# 9891

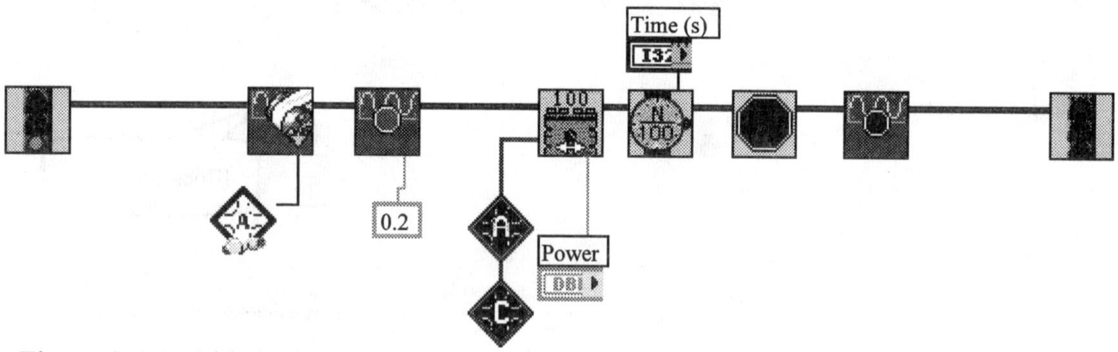

Figure 3.4.4. This is the ROBOLAB program for the front panel of Figure 3.26. These icons are placed in the program's "block diagram" while user input and output is managed in the program's "front panel".

Make our example front panel

In the following section, we explain how to create the front panel from Figure 3.4.1. This panel is designed to be used with a single motor car that has an angle sensor. NXT motors have a built in angle sensor. RCX systems will need to mount a separate rotation

---

[4] Currently, MINDSTORMS software does not allow you to create front panels. Customization of the MINDSTORMS software requires LabVIEW and the NXT Toolkit.

[5] Vi stands for virtual instrument. In many modern scientific and engineering applications, the computer screen and a LabVIEW front panel have taken the place of hardwired knobs, switches, and displays on traditional instruments.

sensor (Figure 3.4.3). The front panel has two distinct functions. First, it allows for student input. They can change the power level of the motor and travel time for the car. It also allows students to input the diameter of the wheel attached to the motor. This allows the program to correctly scale the distance graph, changing rotation data into linear distance data. The 2nd function of the front panel is to display data that comes back from the car after it is has been driven.

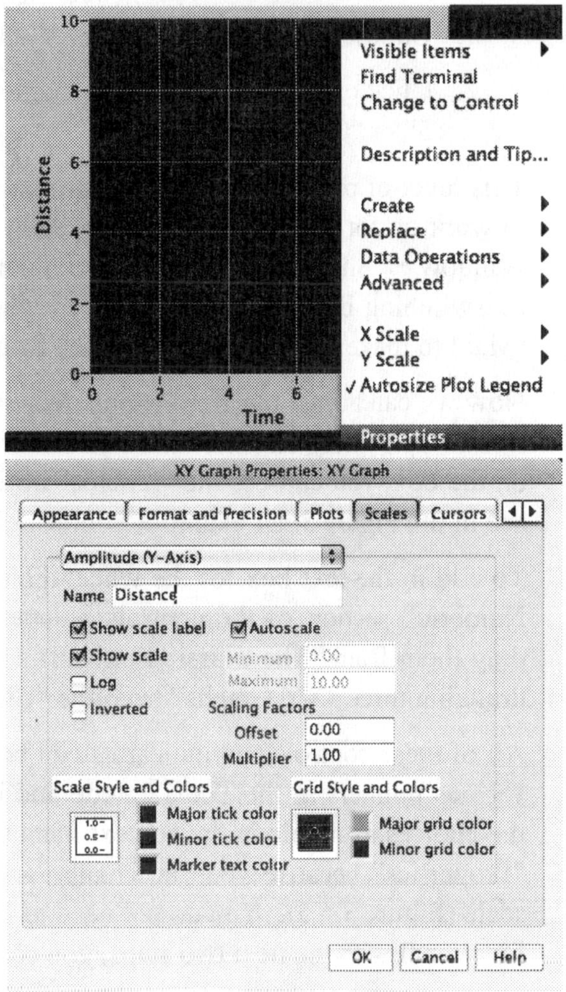

Figure 3.4.5. By right-clicking on an element in a front panel, you can configure the element and change its properties.

We divide our example into five steps.

<u>Step 1: The "Download" Program</u>

Start ROBOLAB and navigate to Inventor 4. In your block diagram window, write a program to initialize and begin data logging a rotation sensor. Add a motor to drive the car forward at an inputted power level. Enable user input by creating a control (right click on the wiring point of the motor and choose "Control"). The length of time for motion can also be inputted (right click on the wiring point of the "wait for time" icon and choose "Control"). Stop data logging and stop your motor. See Figure 3.4.4.

Step 2: Creating the Front Panel Objects

For the second part of front panel programming, a higher level of Inventor is required. This is the level that is available through Investigator and is called G CODE. You do not have to switch to Investigator, however. It can be accessed directly from your Inventor program by:

1. Choose "Change Inventor level" in the Projects Menu.
2. Click on the wrench icon to toggle levels and choose either "Programming Level 5" or "Compute Tools 5".

This level of programming allows programming on the Front Panel and this is the place to work on for Step 2. If you do not see your front panel, you can access it by choosing Window → Show Front Panel. A keyboard short cut is handy to know because you will be switching back and forth between windows often. Use ctrl-E (PC) or command-E (Mac) to make the switch between the front panel and block diagram windows.

Now we can adjust our front panel. When you created the two controls in your program from Step 1, they showed up on your front panels as numeric boxes. If you right-click on the box, you can use the "replace" option to change them to the slider and meter you see in the Figure 3.4.1 example front panel.

To add in the text box for the wheel diameter, drag a numeric box in from the "Classic Numeric" section of the controls palette. To add the Download and Upload buttons, drag them from "The Classic Boolean" section of the palette. Next, add your graphs by dragging three "XY Graphs" from the "Classic Graph" section of the controls palette.

All of these controls and indicators can be resized, relabeled, and formatted by using the various pointers in the Tools Palette and by right clicking on the objects tinkering with the choices available in the context menus. One extremely useful context menu option is "Properties" because you can change a graph's data range, accuracy, labels and axes scale (Figure 3.4.5). You do not need to finalize your front panel now, once you make it functional with the next two steps, you can go back and make final adjustments.

Step 3: Setting up Events

The next step involves separating the download program from the upload and analysis (GCODE) program. These two separate tasks are called events and they are controlled by the "Download" and "Upload" buttons you put into the front panel. In the download event, students send their motor power and time commands to the vehicle. In the upload event, students tell the NXT or RCX to send its data to the computer so it can be displayed in the three graphs. To establish the download and upload events, go back to the block diagram window.

You will notice that your block diagram has a few new objects in the window. Each one represents modifiers matching the controls and indicators from the front panel. Also, since you changed your Inventor level, the block diagram functions palette for

Programming Level 5 now contains options for "Data Logging And Motors", which is the traditional palette for Inventor 4, and a G CODE palette where you will obtain the data analysis and LabVIEW programming structures necessary for this step.

Click on the G CODE palette to define your download and upload events. To do this, you will need to find the "event structure" in the G CODE palette (Figure 3.4.6). Click on it and drag it around your ROBOLAB code.

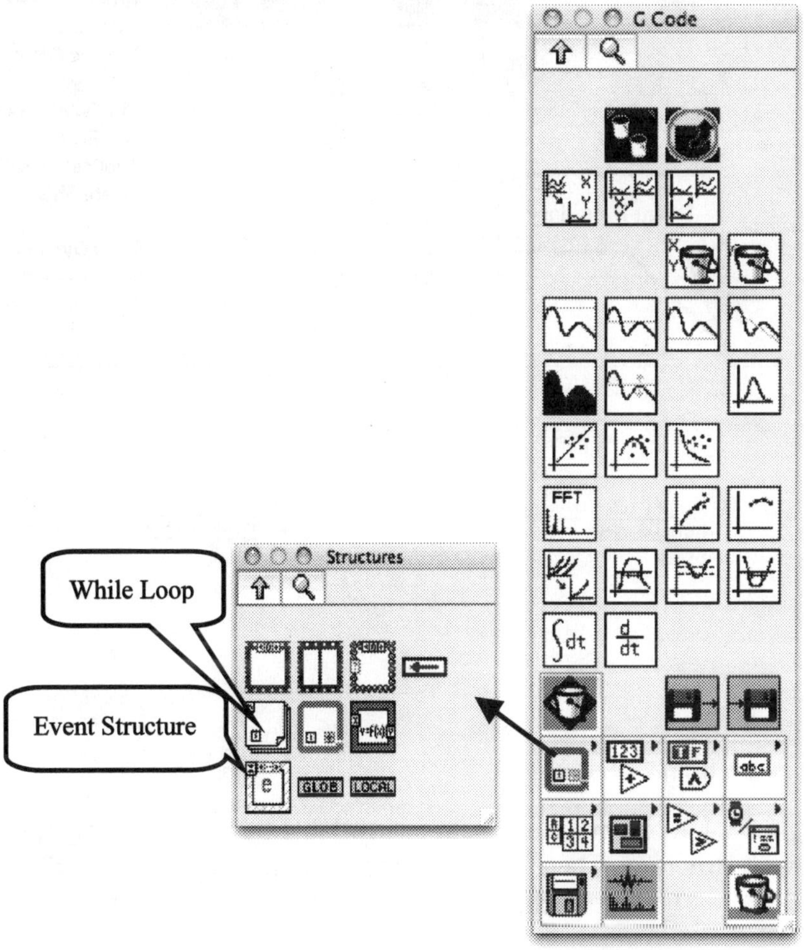

Figure 3.4.6. The G Code palette. You will need Event Structure and the While Loop to create the example program seen in Figure 3.4.7.

These event structures are powerful because they wait for an input from the user. In this case, the input is a click on the "Download Now" button. To include the "Download Now" button, you need to drag the "Download Now" Boolean input (the T/F) into the event structure. Right-click on the "Timeout" label at the top of the event structure. Choose "Edit Events Handled by This Case". This brings up an "Edit Events" window (Figure 3.4.7). Choose "Download Now" and "Value Change". This means that when the "Download" button is toggled on and off the program in your event structure will be downloaded to the NXT. Finally, wire the "Download Now" Boolean object to the edge of the event structure as shown below in Figure 3.4.8.

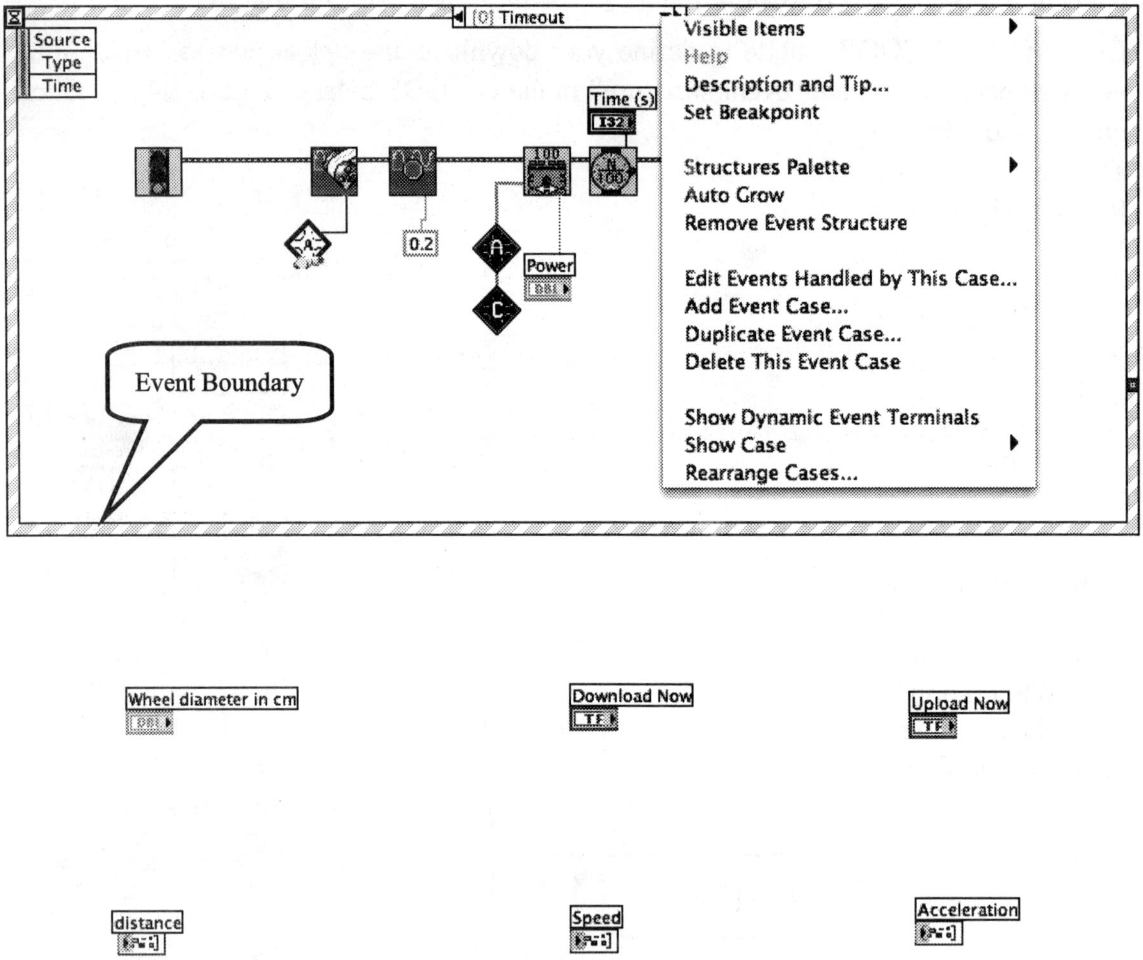

Figure 3.4.7. To make an event controllable by a button, drag the button's identifier into the event structure. Then, configure the event by "right-clicking" on the top bar of the event. Choose "Edit Events Handled by This Case" from the context menu.

<u>Step 4: The UPLOAD program</u>

First, add a second "Event Case" to the existing event structure by "right-clicking" on the Structure boundary. Choose "Add Event Case". Drag your "Upload Now" Boolean object into the event. Repeat the procedure from the previous step to edit events. Choose "Upload Now" and "Value Change". Click OK and wire your "Upload Now" Boolean object to the edge of the Structure (Figure 3.4.9)

You now have two cases in your event structure. You can switch between these by clicking on the arrows in the case label at the top of the structure.

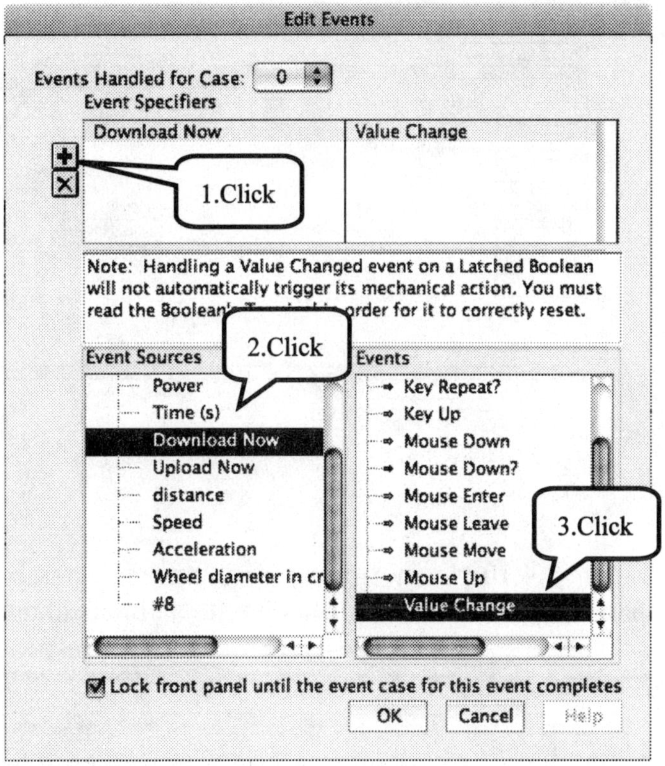

Figure 3.4.8. The Edit Events window allows you to control what triggers the event. In the case of the "Download Now" button, we are looking for a "Value Change". This occurs when the button is pressed. Its value changes from a False (not pressed) to a True (pressed).

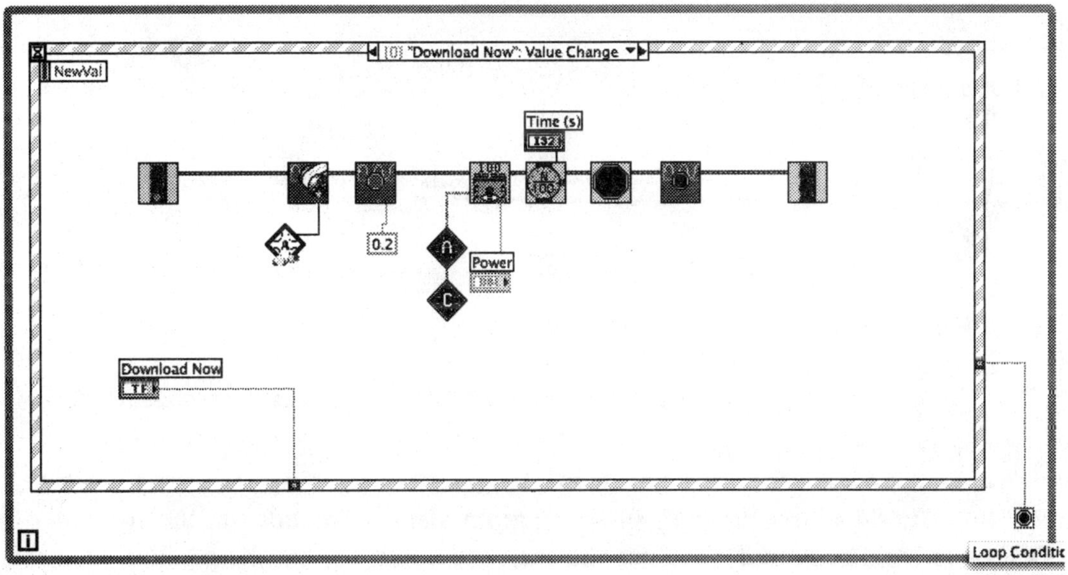

Figure 3.4.9. Wire the "Download Now" Boolean to the edge of the event structure as shown here. Note: the while loop structure placed around the event structure will be covered in the last step. It will keep the front panel running until the "Stop" button is pressed.

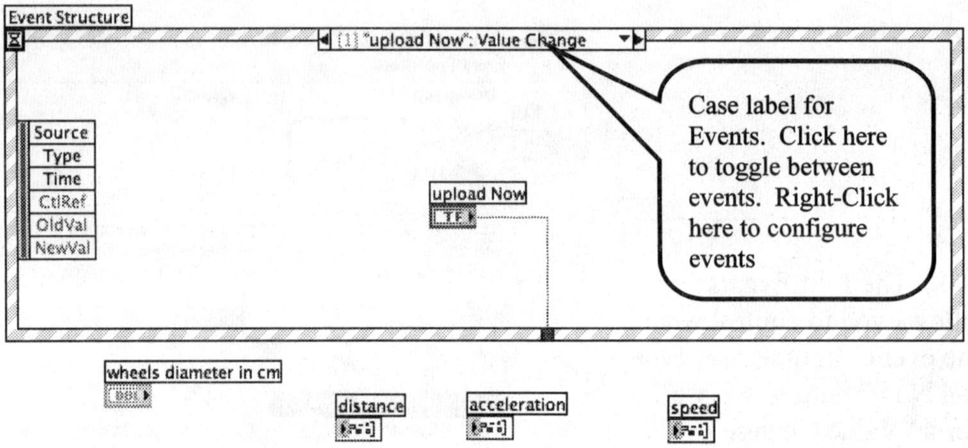

Figure 3.4.10. Event structures can have any number of cases. If you think of your front panel as a computer application with a graphical user interface, the cases within an event structure allow you to respond to user input.

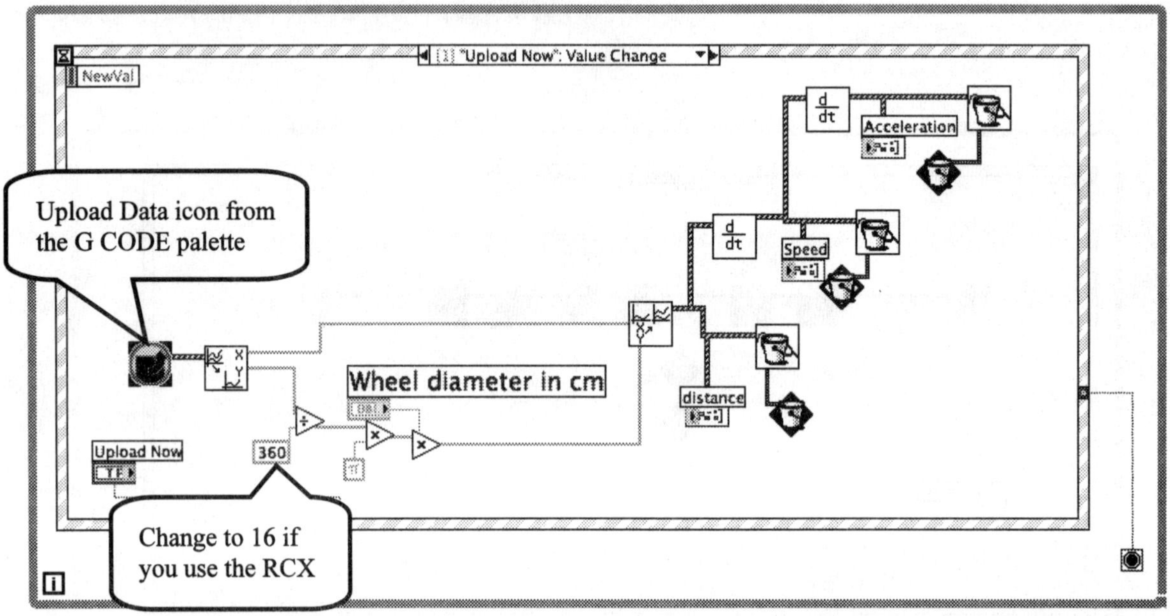

Figure 3.4.11. In the above block diagram program, data is uploaded to the computer from the NXT or RCX. The X and Y values are then separated so that the rotation data can be converted to linear distance values. For the NXT motor encoders, a value of 360 from the sensor equals one rotation. For RCX angle sensors, a value of 16 from the sensor equals one rotation. After this conversion, the X and Y values are put together, plotted in a distance vs. time graph and then differentiated twice to generate speed vs. time and acceleration vs. time graphs. The bin plot icons shown after each graph icon enable the user to store the data for use in further analysis if necessary.

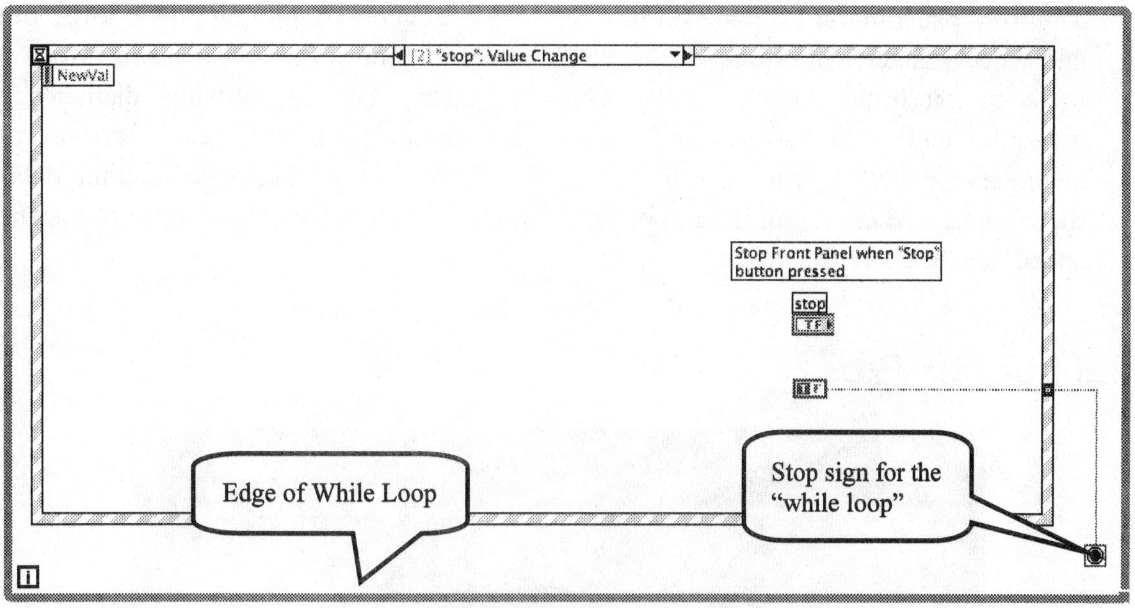

Figure 3.4.12. In the final step, a "while loop" is added around the case structure to keep the front panel's program running until the "Stop" button is pressed. This "Stop" is an event case that needs to be added to the event structure. Like the previous steps, this case looks for a value change. When that change occurs, a value of "true" is passed to the while loop's stop sign. This will end the loop and stop the front panel's program.

In the "Upload Now" event window, use icons from the G Code palette to build the program that analyzes and presents the uploaded data. See Figure 3.4.10 and Figure 3.4.11 for more details.

Step 5: Keep the Front Panel's Program Running Until the "Stop" is Pressed

This final step is not essential. It is more of a convenience. In this step, you need to add a "while loop" around your entire case structure. Next, create a "Stop" button in your front panel, return to the block diagram and drag it's Boolean identifier into the new case. Add another case to the event structure and choose the Stop button as the case specifier. Drag the Stop button's Boolean identifier into the event structure and wire a boolean T/F into the stop sign of the while loop (Figure 3.4.12). The T/F icon is found in the G CODE palette. When the program is run, the loop will keep the front panel running until a value of "true" is passed to the loop's stop sign. This happens when the "Stop" button is pressed.

That's It! The "Download" and "Upload" programs are now hidden while the user interacts with the front panel. In other words, the user does not need to know how these programs operate in order to use the front panel to do some very powerful physics projects!

Front panel operation

The front panel in our example (Figure 3.4.1) is operated by inputting power level, time, and wheel diameter information. Move the slider to control the power level. Move the meter to control the duration of the vehicle's motion. Type in the wheel diameter and then click on the "Download Now" button. Run the program on the car. Reconnect (if necessary), the NXT or RCX and click on the "Upload Now" button to send the data to the computer where it will it will be displayed on your front panel as a series of distance, speed, and acceleration graphs.

3.5 Logger *Pro*®

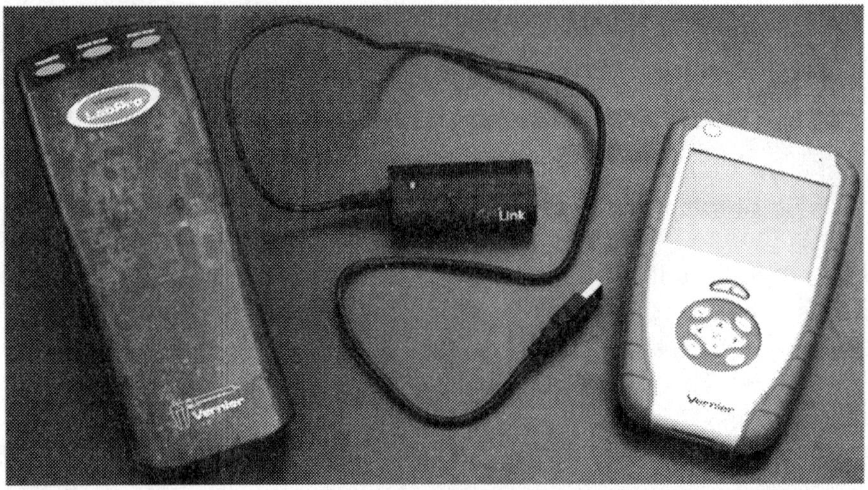

Figure 3.5.1. When using Logger *Pro*, you need to connect sensors to Vernier interfaces such as the LabPro®, Go!™Link, or LabQuest.

Background

Vernier Software and Technology has been making data logging tools for teachers since the 1970s. Starting in 2006, Vernier, LEGO, National Instruments, and the Tufts Center for Engineering Education and Outreach (CEEO) formed a partnership in an effort to integrate the robotics and data logging technologies offered by the four organizations. In this section, we present to you a guide for using Vernier's Logger *Pro* software in robotics projects. One key difference between data logging via MINDSTORMS or ROBOLAB and Logger *Pro* is that the NXT is not used as the interface between the sensor and the computer. You will need to connect the sensor to the computer through an interface manufactured by Vernier (Figure 3.5.1). If you are using sensors from other companies such as Pasco, Fourier, or Onset, use interfaces that are compatible with their software. While we do not present any additional information specific to software from these companies, it is likely that the features we highlight here are easy to find in their data logging programs.

While data logging software from Vernier, Pasco, Fourier, or Onset does not directly interface with the robot and requires additional hardware, we present this option to you because many teachers and students have years of experience with the software and sensors from the companies mentioned above. We encourage you to still use these tools in your robotics projects. The inconvenience of adding a separate interface to a project is offset by the familiarity you have with the software. Additionally, we chose to include Logger *Pro* in the book because it is a highly refined product with easy to use data collection and analysis features for beginner and advanced users. Whether you are new to data logging software, or a veteran user, software like Logger *Pro* is a valuable tool in your "Physics with Robotics" projects!

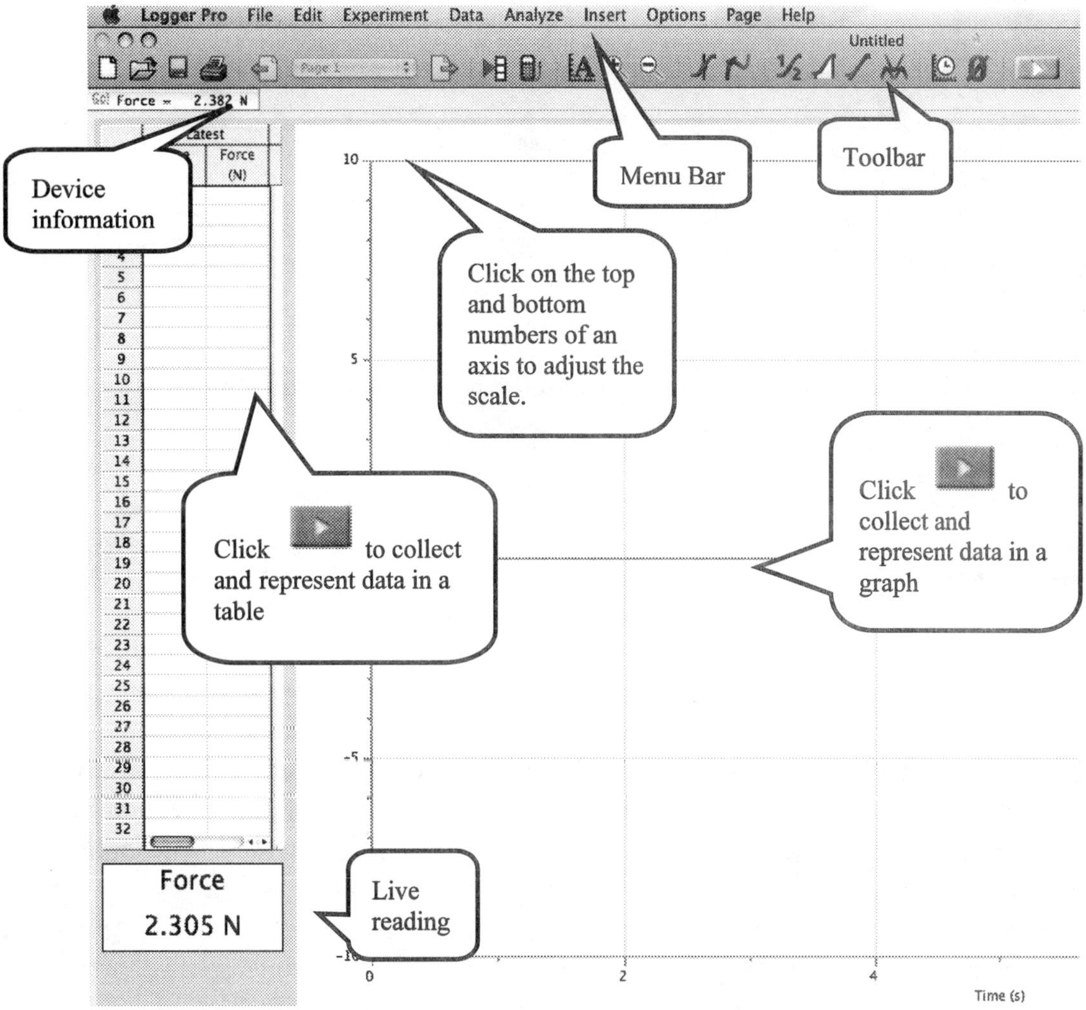

Figure 3.5.2. Logger *Pro* window.

Getting started

To start using Logger Pro, you can either start from scratch by opening the program itself or by double clicking on a previously made file. You can also access previously made files by using the "File → Open" menu item. If you have a sensor connected to the computer directly or through a Vernier interface, it will be automatically[6] identified and you can start collecting data. Figure 3.5.2 represents an example Logger *Pro* window.

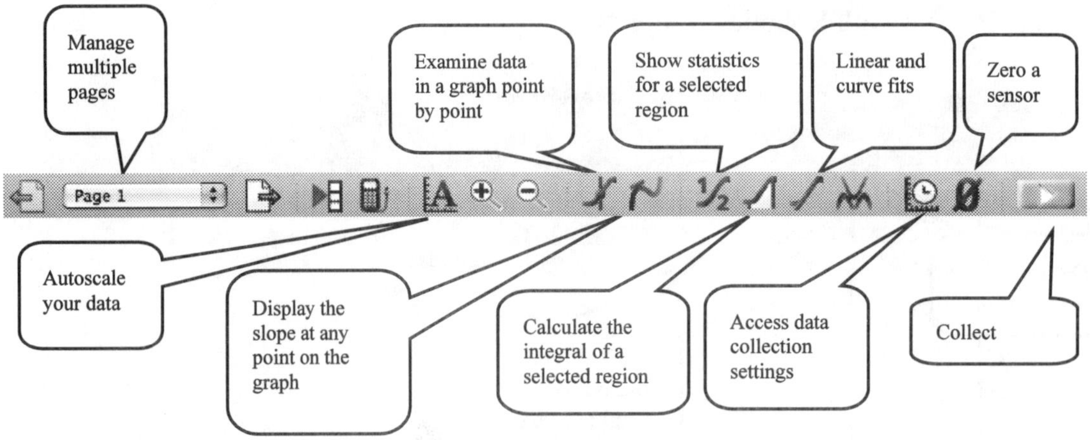

Figure 3.5.3 Provides more information about the icons found in the toolbar.

Figure 3.5.3. Logger *Pro*'s toolbar.

Data logging

Once you have connected your desired sensor, you can confirm that you are ready to collect data by looking at the program's "collect" button. It will be green when the program is ready to collect data. If you do not change any other settings, the program will use a default data collection rate and duration, which is set for each sensor. To meet the needs of your investigation, you will want to click on the data collection settings button in the toolbar or choose the Experiment → Data Collection menu option and make adjustments. Figure 3.5.4 shows the data collection options that you can adjust. To make these settings permanent for a given lab activity, you can save your file and use it as a template before student data is collected. If you do this, keep a spare copy of your "template" file in case students save their project and do not rename the file!

---

[6] Legacy sensors connected via adapters are not automatically identified. They must be manually set-up via the "Experiment → Set-up Sensors" menu item.

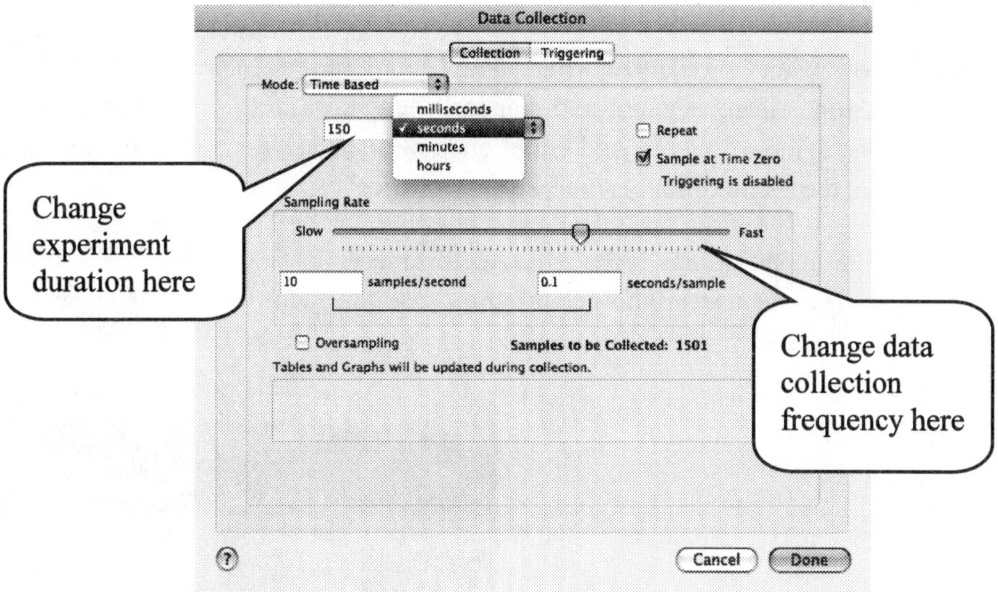

Figure 3.5.4. The Data Collection window allows you to fine tune your experiment.

Example Project – Friction analysis

To provide additional tips for collecting and analyzing data, we provide here an example project taken from Chapter 5. For this example project, we will analyze the force of friction as a LEGO MINDSORMS vehicle pulls a LEGO sled carrying a Vernier Dual Range Force sensor across a lab counter. For this example, we built a single motor NXT car and attached it to the sled (Figure 3.5.5). The single motor RCX car depicted in Chapter 2 would work equally well. We used the following steps to collect and analyze data to determine the force of static and kinetic friction between the sled base and the lab bench:

1. Using MINDSTORMS NXT software or ROBOLAB, program the car to move forward at a constant speed for 4 seconds.

2. Tie the sensor hook to the robot and push the robot until the string is taut. Click on the "zero" button to zero the force sensor.

3. Press the collect button and then run the robot so that it pulls the sled.

4. Inspect your data. If you need to redo the run, set it up again and press the collect button. Logger *Pro* automatically erases over the latest run. If you want to save a run while collecting additional data then use the menu option "Experiment → Store Latest". If you are managing several runs of data, click

on the Y-Axis labels to configure which data sets you want to show in your graph (Figure 3.5.6).

5. If you want to analyze your data to determine basic statistics such as max/min, mean, median, and standard deviation, drag your mouse over the graph region of interest and either click on the statistics button on the tool bar or use the menu option "Analyze → Statistics".

6. Use the menu option "File → Save" to save your data. If you wish to export your data for use in another program, use the menu option "File → Export as".

Figure 3.5.5 Single motor NXT vehicle pulling a force sensor that is connected to the computer using the GO! Link interface (not shown in picture).

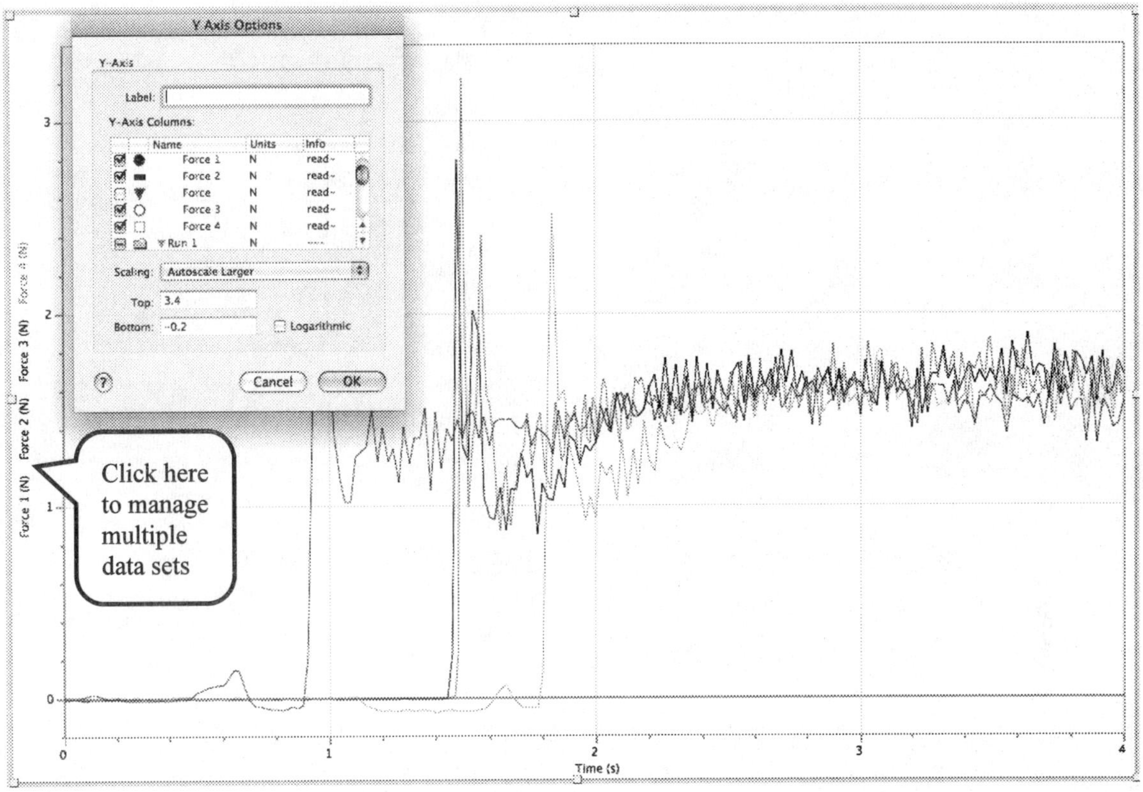

Figure 3.5.6. If you need to manage multiple data sets, click on the Y-Axis labels. This brings up a Y-Axis Options window. Use this window to choose data sets will be displayed.

There are many other features in Logger Pro not demonstrated in this quick example but these six steps will get you started using Logger Pro for data logging in the activities of this book. For a more complete tutorial on how to use Logger Pro, visit http://www2.vernier.com/manuals/LP3QuickRefManual.pdf.

Chapter 4

Motion

4.1 Going the Distance

Age Group: 14+

Essential Question: Can you tell a robot how long to travel an *exact* distance?

Recommended Time: 45 to 90 minutes

Materials:

NXT or RCX Kit
Meter stick or measuring tape
LEGO Figures or other miniature characters

Activity Overview:

In this activity, students will be given a set distance between two lines of tape. On one line, students will place their LEGO vehicle. On the 2nd line, a row of LEGO or other miniature characters will stand. The goal is to predict the exact time to program the LEGO vehicle such that it will travel and stop as close as possible to the miniature character without touching it. This is not a "trial and error" process. Students will complete a series of measurements and organize their data into a model (i.e. graph, data table, and/or equation) that allows them to accurately predict the time required to travel the challenge distance.

Learning Objectives:

By the end of this activity, students will be able to:

1. Create a data table, graph, and/or an equation based mathematical model of an object's motion.
2. Use a mathematical model to predict the motion of an object
3. Test a prediction with an experiment
4. Refine their models based on new experimental evidence

Pre-Activity Discussion Questions:

1. In science or engineering, what is a model?

2. What are some examples where table, graphs, and equations are used as models?

Activity Instructions:

1. Create a basic LEGO vehicle that drives straight or use one built for a previous activity. For the purposes of this activity, "straight" means that your vehicle can drive approximately 3 meters while moving left/right less than 0.2 meters.

2. Using Mindstorms or Robolab software, program your vehicle to drive forward for a certain amount of time. Observe how far your vehicle moves.

3. Consider the challenge posed to you by your instructor – "Figure out the necessary measurements and process by which you will organize those measurements (via equations, graphs, data tables, etc) to predict the time your robot will travel any given distance." In other words, create a model that allows you to predict the time of travel for any distance your vehicle moves.

4. Once you have figured out your model, test your prediction with a series of experiments.

5. Refine your model as necessary so you can get as close as possible to the distance defined in the final "Going the Distance" challenge.

6. When your instructor has created the final "Going the Distance" challenge, do the necessary measurements to predict the time of travel. NOTE – You will not complete this task via trial and error. You only have one chance to meet the "Going the Distance" Challenge.

7. Observe how close you come to the line.

Student Observations and Analysis:

Data

Time of Travel	Distance Traveled	Notes

Draw a graph on graph paper of Distance Traveled versus Time.

Reflection Questions:

1. How close did your vehicle get to the challenge distance?

2. When you organized your measurements for this activity, you created a model that represents the motion of your vehicle. Your model may have been a data table, graph, and/or an equation. What type of model did you use to represent the motion of your vehicle and how well did it work to help you make accurate predictions?

3. If you could do this activity over again, explain what changes you would make (if any) to your LEGO vehicle.

4. If you could do this activity over again, explain what changes you would make to collecting and organizing your data.

5. If you could do this activity over again, explain what changes you would make to the process you used to make your predictions.

6. Think about the steps you followed during this activity. Do scientists and engineers perform similar steps in their work? Think of an example and explain.

7. For your car, can you give a general rule for how far this car goes per second, (speed)? Use this answer to determine the following:
   a. Where would the car be after 10s?

   b. How long would this car take to travel 1m?

Notes and Extensions:

1. While the original design of this activity ends after one challenge distance, consider conducting three, allowing the students to continue refining their models.

2. This activity works very well as one of the introductory activities in a physics curriculum. It lends itself well to follow-up discussions about models – their creation, use, and refinement. As a kinematics activity, it allows students to explore a prediction activity based on their understanding of motion concepts.

3. The Chapter 7 activity – "Time for Robot Golf" works well as a follow-up to or replacement of this activity depending on the goals of the instructor and performance of the students.

Acknowledgements:

The Tufts Center for Engineering Educational Outreach created the original version of this activity. Navigate to http://www.LEGOEngineering.com for the original "Going the Distance" and more activities.

4.2 An Investigation of NXT Motor Speed

Age group: 15+

Essential Questions:
How exact is the motion of an NXT?
Is power level 100, one hundred times faster than power level 1?

Recommended Time: 40 – 50 minutes

Materials:

NXT LEGO kits
NXT data logging software

Activity Overview:

To investigate the NXT motor speed versus power level, we could measure distance traveled for one second, for each power level and convert to speed. Gather the data for all power levels and look for a relationship. This would be an arduous task that could be relieved by sharing the job across the class.

For a more independent activity, why not use the power of programming and rotation sensor data logging and run the motors at all speeds for a constant length of time each. Of course the longer it runs, the more points you have and the more accurate the average speed for that interval.

Learning objectives:

1. Design and perform controlled investigations.
2. Refine investigations after evaluating variations and inconsistencies.
3. Analyzing linear mathematical relationships between variables and determine gradient.
4. Use mathematical ratios to verify physical laws.

Building Ideas:

Any NXT car, one motor is sufficient. See Chapter 2 for examples.

Activity Instructions:

1. You need to increment the power levels from 1 to 100 and run the car for equal length times for each power level.
2. The motor encoder needs to be data logged but not just the total angle, instead calculate the distance moved for each power level. For example run the car through all power levels for a total of 10 seconds. For 100 power levels this means 0.1s per

level. To get a good reading for speed the angle position on the encoder needs to be sampled often e.g. every 0.01s (10 times per level).

3. Use variables or containers to store the angle difference (displacement angle) and the power level so these can be logged. You'll need more than one variable to read angle before and after each level. The difference between these called the displacement angle can then be found.
4. You must still initialize the data logging of the motor encoder and variables of displacement angle and power level. The data logging sampling rate for variables needs to be only once per level e.g. 0.1s.
5. Download the program and run the car on an obstacle free flat course.
6. Upload the data.

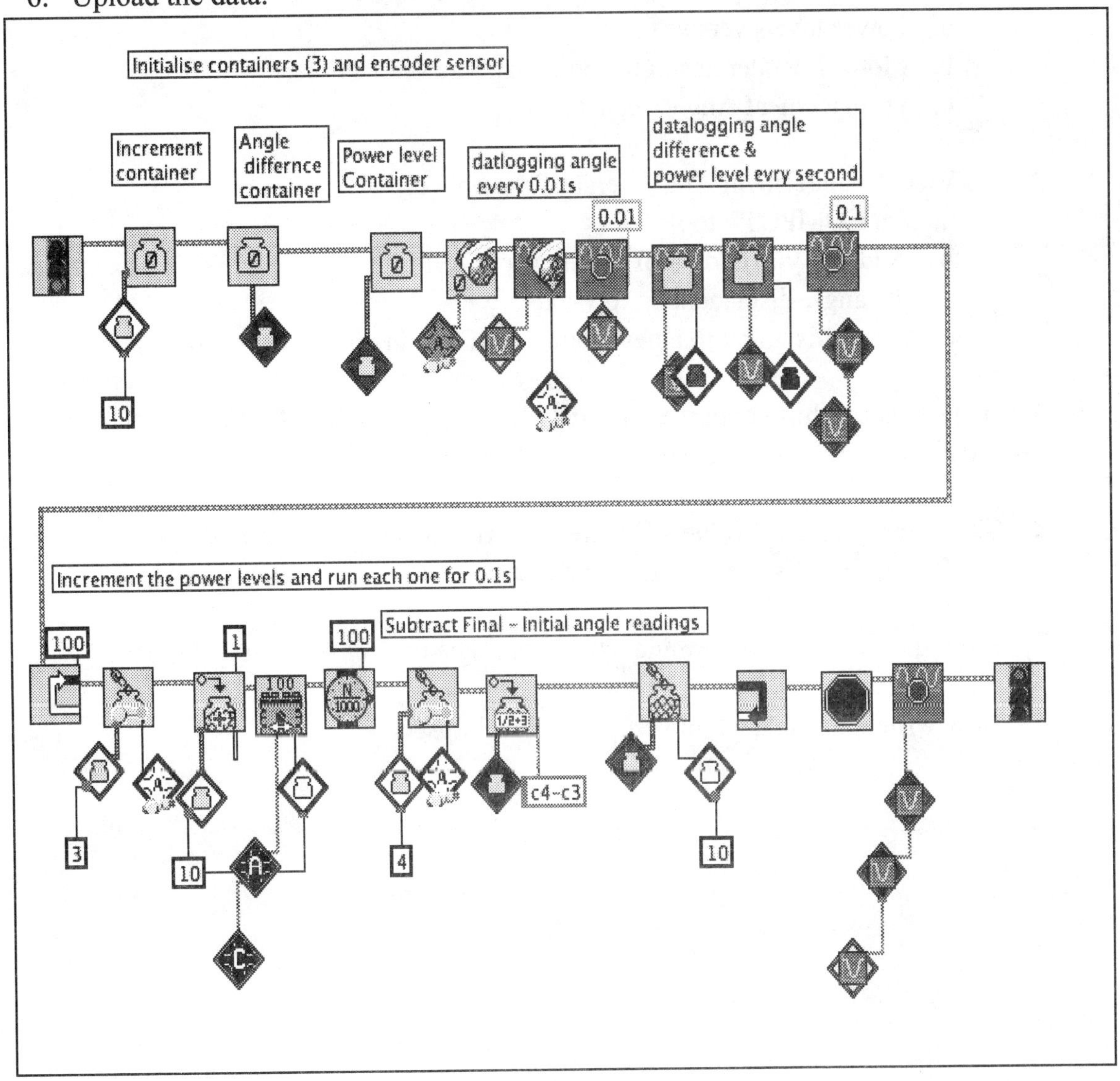

Figure 4.2.1 Sample Robolab program for NXT Motor Speed

Student Observations and Analysis:

1. In your graphical analysis program, you need power level changes on the X-axis and angle distance on the Y-axis.
2. What was the general trend of your data?
3. If the graph was linear, determine the slope.
4. What does this slope represent and how can you use it to determine speed (cm/s) per power level?

Analysis Procedures using ROBOLAB INVESTIGATOR:

1. On the UPLOAD area, 3 data sets should be obtained:
    a. Power levels versus Time.
    b. Motor Encoder absolute angle readings versus time.
    c. Displacement Angle versus time.

2. To View Displacement Angle versus Power level:
    a. On COMPUTE tools 2
    b. Add the y-axis data of the power level to the x-axis and the y-axis data of the angle displacement to y-axis.
    c. Store this graph in a new container. Examine the shape of the graph.

3. If the relationship is approx. linear (which it should be!) find the slope using the VIEW area and Measure window and slope tool.

4. The slope gives the value of degrees of rotation in 0.1s intervals per unit power level. Convert this to rotations/s and also revolutions/s.

5. From the wheel size determine what is the speed factor per unit power level.

6. Now write a formula to convert power level to speed.

Sample Analysis:

Slope of graph = 0.81°/Power level (for 0.1s)
∴ 0.81°/0.1s = 8.1°/s/ power level
Converting to linear speed:
V = 8.1πd/360 cm/s
= 0.39 cm/s per power level

Now write a formula for speed from power level:
V (cm/s) = 0.39 × PL

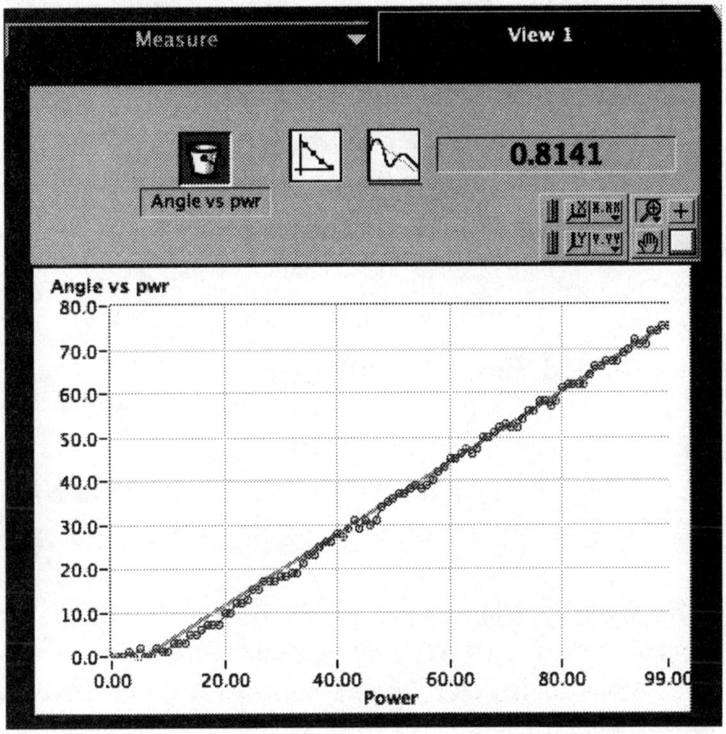

Figure 4.2.2 Sample ROBOLAB Investigator Data

Reflection Questions:

1. Are the NXT motors linear within acceptable error?
2. Explain in a few lines what this means in terms of distance traveled at different power levels.
3. Explain how you would program an NXT to go 1metre at 20cm/s.

Notes and Extensions

1. Verify your formula works by testing your car at various speeds for a certain time and checking the distances. You can upload your data and view using Investigator or the Program from Activity 4.3.

2. Make up a distance – time graph for your partner involving 3 steps with varying speeds (positive, stop and negative for varying times) that they have to convert to power levels and times for their NXT to match! – See Chapter 7: NXT Distance Match Challenge.

4.3 Testing Speed versus Acceleration of Drag Cars.

Age Group: 12+

Essential Question:

In a Drag car race what is more important; high speed or high acceleration?

Recommended Time: 40 – 50 minutes.

Materials:

LEGO Kits (RCX with motor and rotation sensor or NXT with motor/encoder).
ROBOLAB

Activity Overview:

This activity allows you to directly measure distance, speed and acceleration of a car, by displaying graphs on the front panel of a ROBOLAB program. The student inputs the power level, time, and diameter of the wheels, downloads the program, runs the car and is able to upload the data collected back to the computer for analysis.

Learning Objectives:

Compare various factors that may improve speed and acceleration including power, time of travel, gear ratio and wheel size.

Building Ideas:

A simple single motor, geared LEGO Car (RCX or NXT) with a rotation sensor on the front wheel axle.

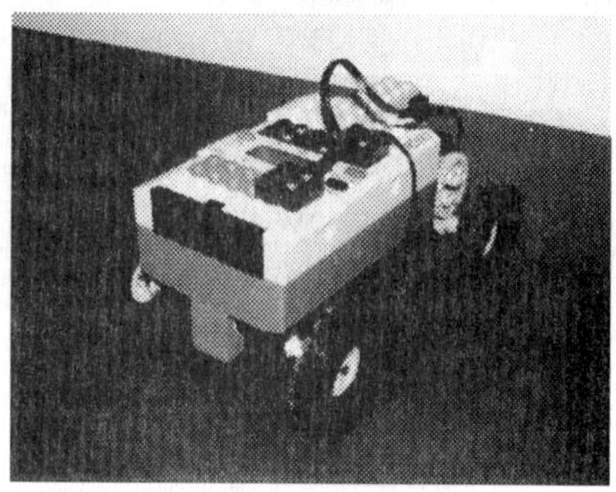

Figure 4.3.1 Simple car /photo

Figure 4.3.2 Simple car/ underside

Figure 4.3.3 Simple car/ add motor

Programming Ideas:

Graphs may be obtained from angle sensor data using Mindstorms 2.0 and ROBOLAB Investigator but for the best comparisons we recommend using a custom front panel run in ROBOLAB. A download of the program, "accelerations.VI" for NXT or RCX is available at our website: http://www.physicswithrobotics.com/ or you can write the program yourself and adapt it how you wish. For more details on how to program front panels see chapter 3, section 3.4. A LabVIEW version of the program will be made available in late 2009.

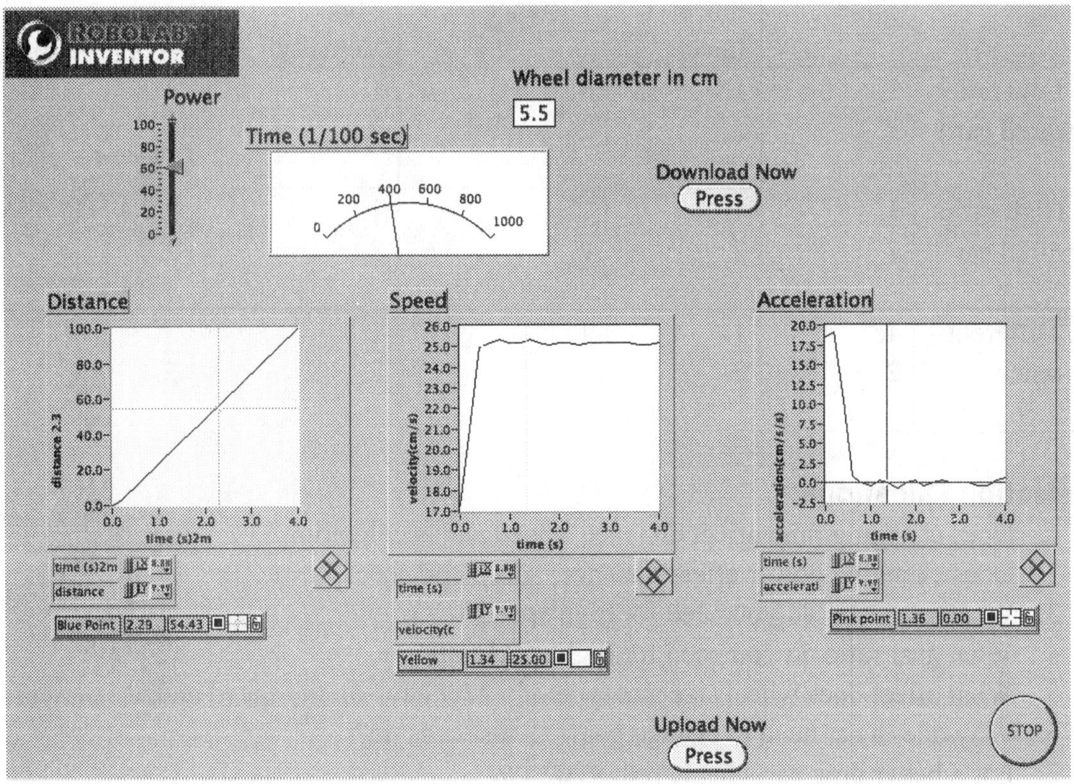

Figure 4.3.4. ROBOLAB Front panel for the Chapter 4 activity: *Testing Speed and Acceleration of Drag Cars.* With custom front panels, all the programming and data logging controls can be placed in one user interface. You can download either RCX or NXT versions of this front panel at our website.

Activity Instructions:

1. Use the Front Panel of the program "accelerations.vi" to input time, power level and rotation wheel's diameter for the download program.
2. Put the car in front of the tower
   Click on white arrow to run and then Press DOWNLOAD Now.
3. Take the car and run it across the room.

4. Put the car in front of the tower. Click on the white arrow (if necessary) and then press UPLOAD Now.
5. Determine the total distance, maximum speed and acceleration. You can use the cross hair pointer and read the y values in the window below the graph.
6. Repeat using a different time and power level. Record results.
7. Try changing the gears around.
8. Try different size wheels on front or back (remember to change the input)

Student Observations and Analysis:

Changes	Time	Power Level	Distance	Max. Velocity	Max. Acceleration
a. Low ratio geared car	1. 2.	1. 2.	1. 2.	1. 2.	1. 2.
b. High ratio geared car	1. 2.	1. 2.	1. 2.	1. 2.	1. 2.
c. Different wheels	1. 2.	1. 2.	1. 2.	1. 2.	1. 2.

Reflection Questions:

1. Does changing time affect distance, speed and acceleration?
2. Does changing power affect distance, speed and acceleration?
3. What gear ratio do you need for high speed?
4. What gear ratio do you need for high acceleration?
5. What affect does wheel size and type of wheel have on the speed and acceleration?
6. Why does a car need to change gears as it speeds up?
7. For a better drag car what is required?

Notes and Extensions:

1. It's important the students test various gear combinations. It is easy to swap the 8 tooth and 24 tooth gears around on the single RCX car.
2. For different wheels, you must alter the wheels diameter window to the correct value for the wheel connected to the rotation sensor only.
3. Further discussion and challenges relate to finding other ways to increase the cars velocity.
4. Make your own Drag Car according to the optimum results of factors you have tested.

4.4 Projectile Motion

Age group: 14+

Essential Questions:

When a projectile is launched straight up from a car moving along a straight line at constant speed, where does the projectile land?

Recommended Time: 40-50 minutes

Materials:

NXT or RCX Kit
Launcher
(e.g. Technic Cannon 32704 and arrow 32133c01 at www.peeron.com/inv/parts)
Motor for Launcher
Data logging software

Activity Overview:

This activity can be used to introduce the concept of projectile motion in the high school physics curriculum.

A projectile is any object, which once projected continues in motion by its own inertia and is influenced only by the downward force of gravity.

There are a variety of examples of projectiles where horizontal and vertical motion occur:

- An object dropped from a moving object e.g. an airplane
- An object thrown from a cliff.
- An object thrown upwards at an angle.

Students have an opportunity to investigate the principle of constant horizontal velocity of a projectile through building a model car that will launch an object vertically as it moves.

Learning Objectives:

1. Learn about examples of projectile motion
2. Understand vertical and horizontal components of motion
3. Analyze projectile's path by considering the horizontal and vertical components

Building Ideas:

1. Build a simple sturdy LEGO/NXT car
2. Add to the car a simple launcher, such as the one shown using a Spring- Loaded Rubber tipped dart LEGO piece (*see* Figure 4.4.1).

Figure 4.4.1: LEGO Model with launcher

Activity Instructions:

1. Program the car to move with constant velocity (use a constant power level).
2. Program the motor on the launcher to start 4 seconds after the car moves and run for 1 second.

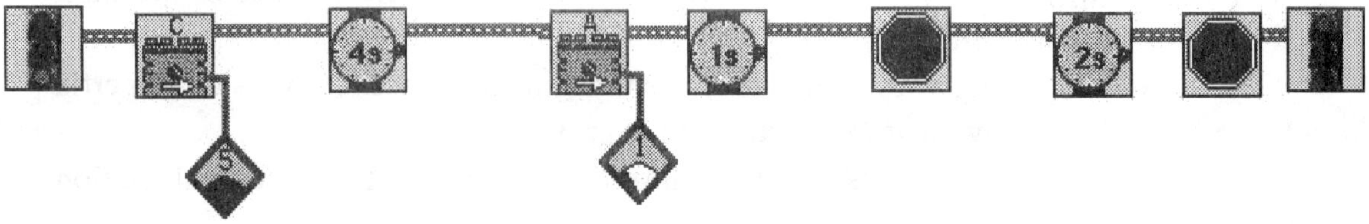

Figure 4.4.2: Sample ROBOLAB program in Inventor 3.

3. Place the object for launching.
4. Test the model launcher without running the program:
    - Manually launch the projectile and test that it lands back on the launcher.
    - If not adjust it for a vertical flight.
5. Start the program and observe the landing as the car moves.
6. Try a different speed for the car and repeat the investigation.

Student Observations and Analysis:

1. Where does the projectile land when it is launched vertically from the moving car?
2. Was it different to the launch from the stationary car?
3. Does higher speed have any effect on its landing?

Reflection Questions:

1. What factors affected the motion of the projectile after it left the car?

2. Can you be sure the car moved at a constant velocity? How can you test this?

3. A car is moving along a straight line at a constant speed of 25 m/s. It launches a projectile straight up. What is the value of the horizontal velocity component of the projectile at the highest point? Neglect air resistance. Explain your answer.

4. Which is true:

   Ideally, a projectile's horizontal velocity component:

   a. Does not change.
   b. Changes most rapidly near the bottom of its trajectory.
   c. Changes at a constant rate.
   d. Changes most rapidly near the top of its trajectory?

   Explain your answer!

5. A car is moving with acceleration along a straight line. It launches a projectile straight up. Where does the projectile land? Explain your answer.

6. A car is decelerating in a straight line. It launches a projectile straight up. Where does the projectile land? Explain your answer

7. A car is moving in a circular path with a constant speed. It launches a projectile straight up. Where do you think the projectile will land? Explain your answer.

Notes and Extensions:

1. Write a program in which the car accelerates while it is launching the projectile. Predict the landing and then investigate it.

2. What happens if the car changes direction while it launches the projectile, for example moves in a circle? Try to investigate this.

4.5 Simple Harmonic Motion

Age Group: 16+

Essential Questions:

What are the characteristics of the motion of a vertical oscillating spring?

Recommended Time: 40-50 minutes

Materials:

NXT kit	Ultrasonic NXT sensor
Retort Stand with clamps	Helical (coil) spring

Activity Overview:

In this experiment the students use an NXT and the ultrasonic sensor to investigate the change in vertical motion of an oscillating spring. Calculations of the amplitude and period can be made using the data from the resulting graph.

Learning Objectives:

Study periodic motion and the characteristics of the displacement graphs.

Pre – Activity Discussion:

1. Introduce the concept of Periodic Motion and define amplitude (A), period (T) and frequency (f =1/T).
2. Discuss characteristics of periodic motion as in example 4.5.1.

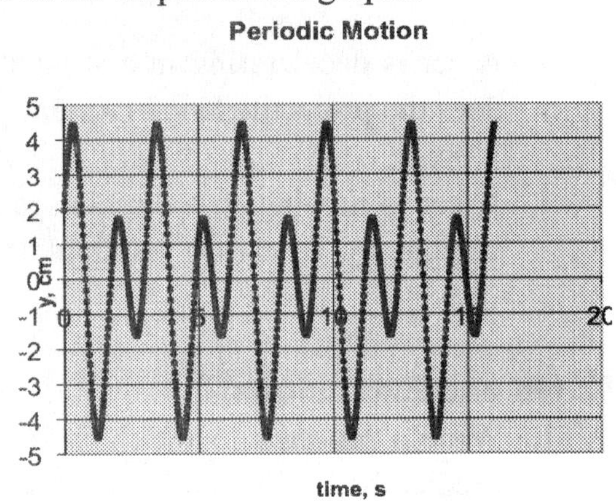

Figure 4.5.1 Periodic Motion

3. Discuss the characteristics of Simple Harmonic Motion as in example 4.5.2. Also explain restoring force is directly proportional to the extension and in the opposite direction (negative).

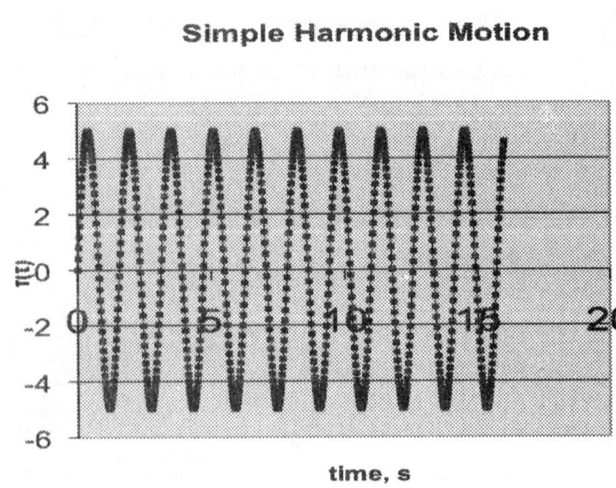

Figure 4.5.2 Simple Harmonic motion.

4. Show students the equation that describes Simple Harmonic Motion:
y = Asin(2πf)t, where A = Amplitude, f = frequency of motion, t = elapsed time (s)

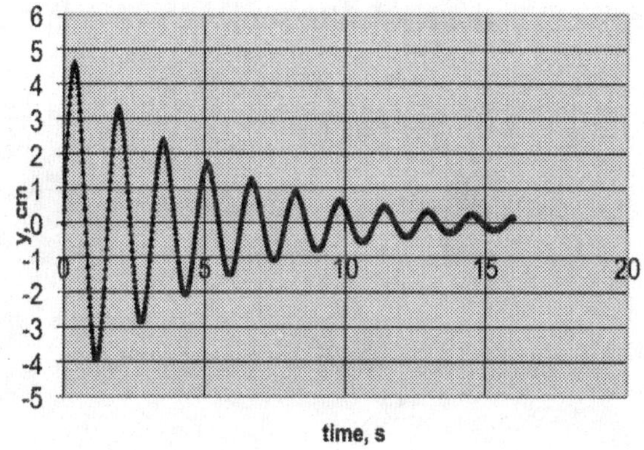

5. Discuss the characteristics of damped motion and what parts of it are periodic.

Figure 4.5.3 damped periodic motion

Building Ideas:

1. Build a model of the spring-mass system using NXT, ultrasonic sensor, spring, and stand. *See* Figures 4.5.4 and 4.5.5
2. Ensure the ultrasonic sensor will read a hard flat surface e.g. wooden floor.

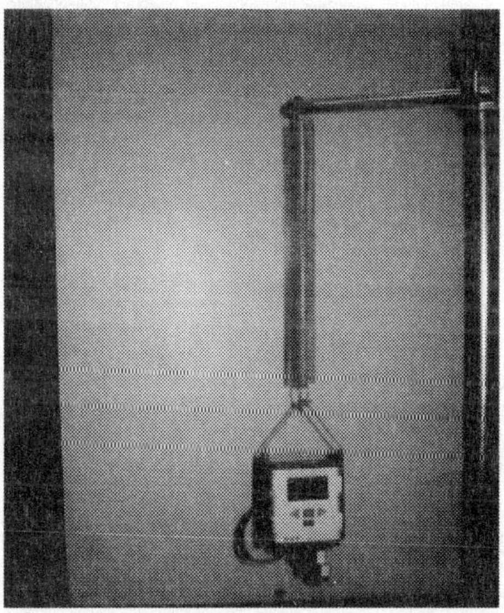

Figure 4.5.4: NXT set-up

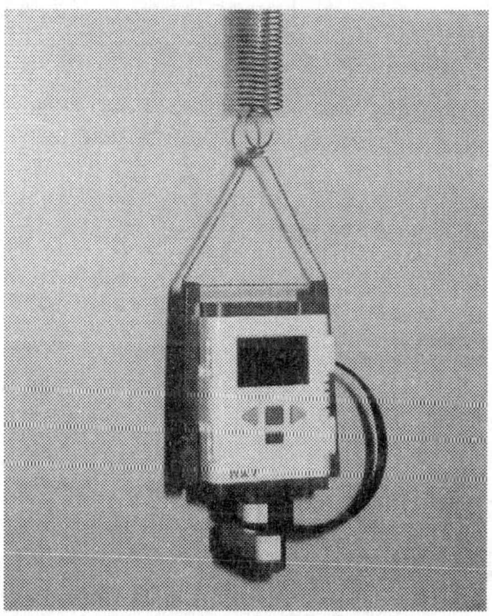

Figure 4.5.5: NXT ultrasound sensor set-up

Activity Instructions:

1. Using Mindstorms 2.0 or ROBOLAB Investigator, write a program to collect the NXT ultrasonic sensor data reading every 0.1s for 10s. See Figure 4.5.6

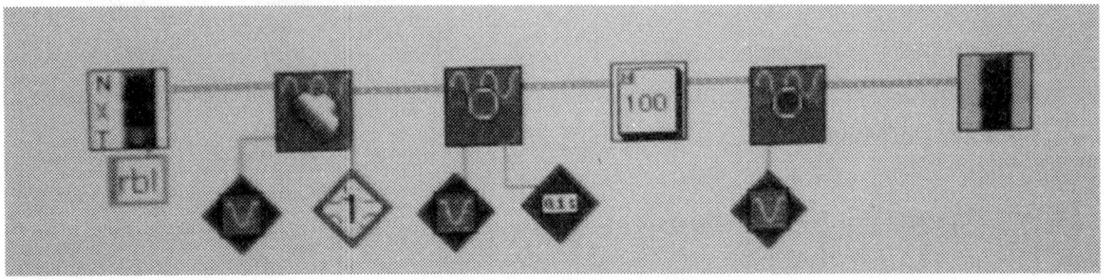

Figure 4.5.6 Robolab program

2. Start the program before you begin the oscillation.
3. Gently set the NXT on the spring to oscillate a few centimeters.
4. After finishing the experiment, upload the collected data.

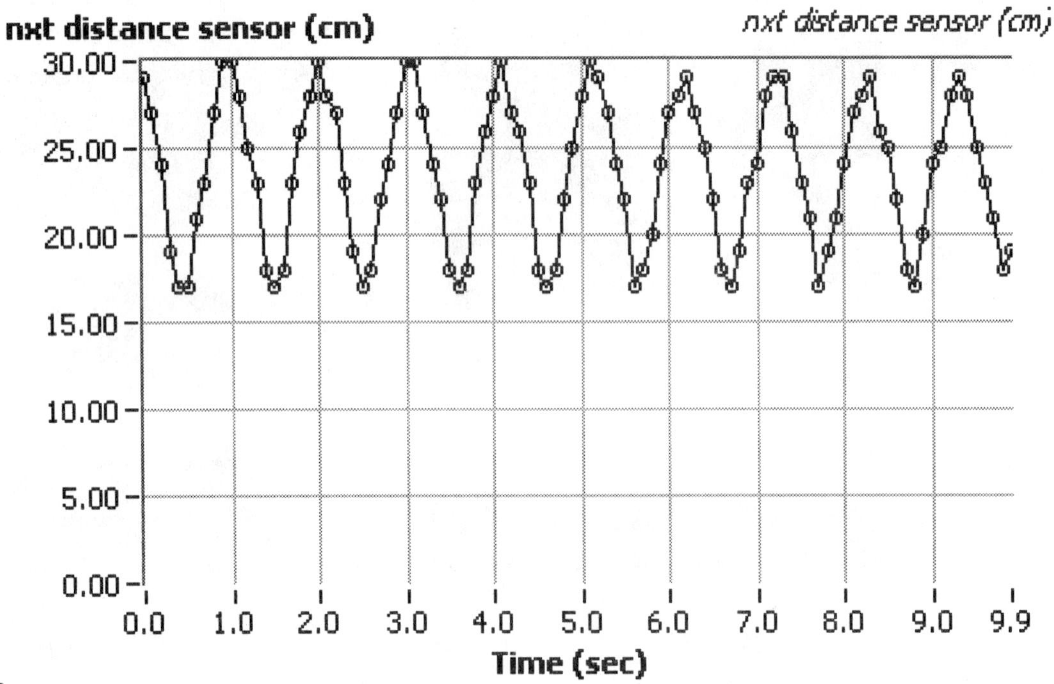

o

Figure 4.5.7 Sample data collected using Investigator

5. Use the experimental data to find the amplitude and period. Calculate its frequency.
6. Repeat the experiment for different amplitudes.

Student Observations and Analysis:

Experiment	Amplitude	Period	Frequency
1.			
2.			
3.			

1. Describe the shape of the graphs.
2. How were they different?
3. For each graph, what changed over time?

Reflection Questions:

1. What factors affected the smoothness of the graph?
2. If the graph showed dampening, what may have caused it?
3. Did increased amplitude have an effect on the dampening?

Notes and Extensions:

1. Use the ROBOLAB Compute Tools 4,5 to perform a Fast Fourier Transform (FFT) on your data set to find frequency content of the data set (*See* Figure 4.5.8). Define the dominant frequency as a natural frequency of the oscillator in this experiment from the FFT graph. Example of this program is shown below. Using this program perform the FFT on your time series data (stored in the red bin) to produce the frequency domain data, which is stored in the green bin in this example program.

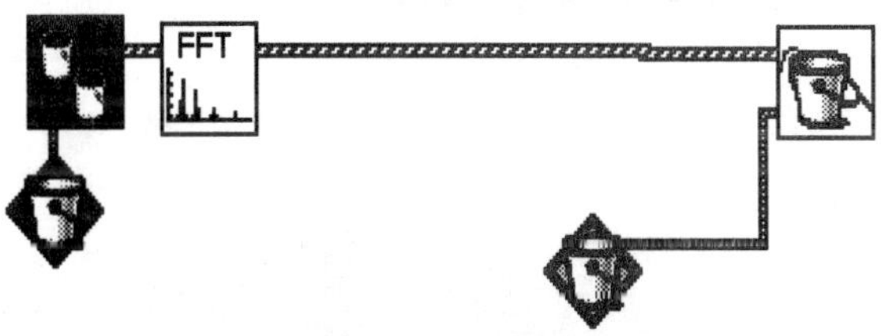

Figure 4.5.8: Program written in Compute Tools to perform the Fast Fourier Transform (FFT) on the data set.

Figure 4.5.9: Results of the Fast Fourier Transform (FFT) on the data set. When result of the FFT is plotted, the natural frequency is evident as the most dominant peak in the graph.

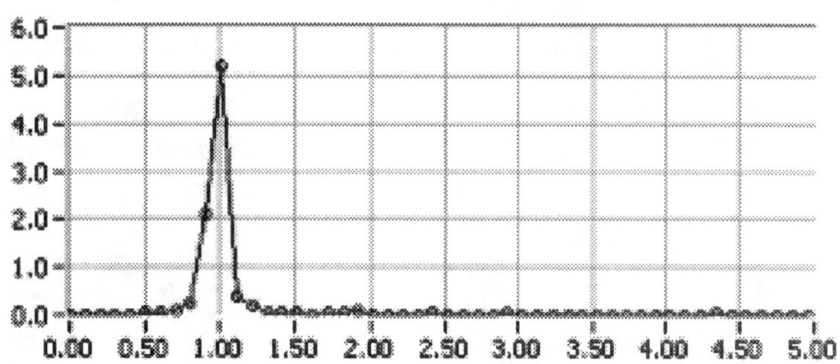

4.6 Swinging with Gravity

Activity Level: 14+ years

Essential Questions:

Can astronauts measure time using a pendulum while they are traveling to Mars?

Recommended Time: 40-50 minutes

Materials:

NXT kit
NXT light sensor
Retort Stand and Clamps String

Activity Overview:

In this experiment a light sensor can be attached to the NXT to collect real-time data on the received signal change as the pendulum swings over a light/dark area. The period of the pendulum can be calculated from the resulting graph.

Learning Objectives:

Examine simple periodic motion.
Use the pendulum formula to compare measured and calculated period.

Activity Instructions:

1. Build the pendulum using some string, the NXT and a light sensor (*see* Figures 4.6.1 and 4.6.2).
2. Use a piece of white paper with a dark bar on it (or a white bar on a dark paper) and place it directly under the pendulum.

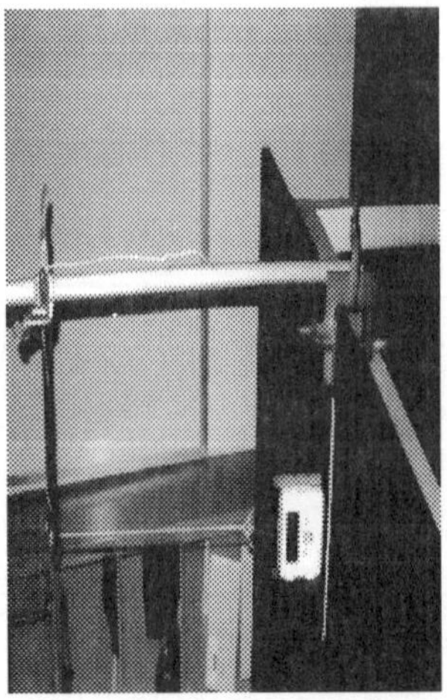

Figure 4.6.1 Experimental set-up

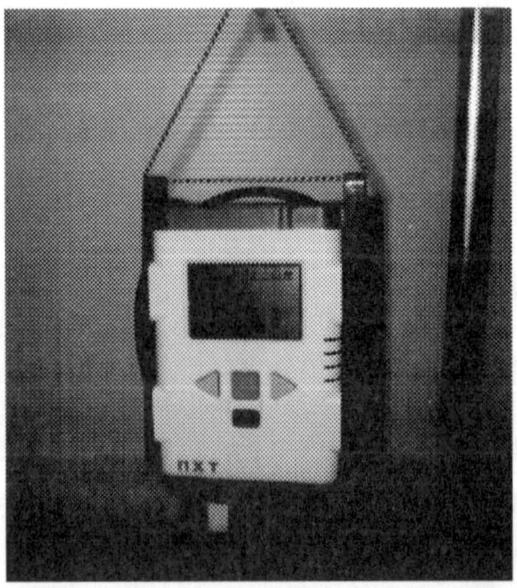

Figure 4.6.2 NXT and light sensor

3. Program the NXT using Mindstorms 2.0 or ROBOLAB to collect light sensor data during for 10 seconds.

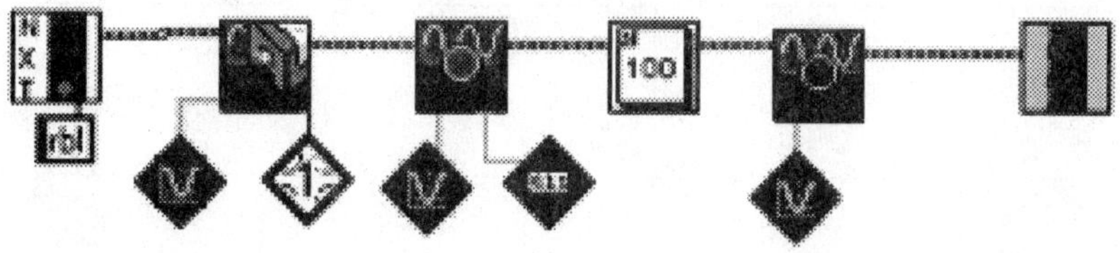

Figure 4.6.3: Sample ROBOLAB program

4. Set the pendulum in motion.

5. Upload the collected data using Mindstorms 2.0 or ROBOLAB.

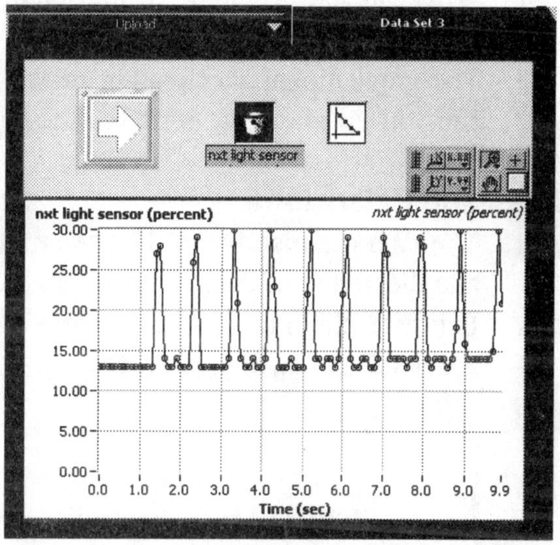

Figure 4.6.4: Sample NXT light sensor data

6. Measure the length of the pendulum (length of the string + half of the NXT length) and record it in Table 1.

7. Calculate the period of the pendulum (T) from the light sensor data graph.
Question: If the period of the pendulum is the time for one complete swing (out and back), what number of peaks will you have to include, to find the time difference?

8. Calculate the expected period (T) using: $T = 2\pi\sqrt{\frac{L}{g}}$

   T is the period of oscillation (seconds)
   L is the length of the pendulum (meters)
   g is the Earth's gravitational acceleration (meters/ second^2)

9. Compare the results for T from Steps 7 and 8.

10. Repeat the experiment but change the length of the pendulum.

Student Observations and Analysis:

Table 1

L (Pendulum length)	$T_{calculated}$	$T_{experimental}$	% Error
1.			
2.			

Reflection Questions:

1. What can you say about the motion of the pendulum? Is it periodic?
2. How does changing pendulum length change the period of oscillation?
3. The gravitational acceleration in the mountains is less then at the sea level. What can you say about the period of the pendulum at high altitudes?

Notes and Extensions:

1. The data can also be used to determine the gravitational acceleration, if the length and the period of the pendulum are measured.
2. Estimate the mass of the Earth using the experimentally obtained value of acceleration due to gravity. Use:

$$g = \frac{Gm_E}{r^2}$$

3. Use the ROBOLAB Compute Tools 4, 5 to perform Fast Fourier Transform (FFT) on your data set to find frequency content of the data set. Define the dominant frequency as a natural frequency of the oscillator in this experiment from the FFT graph. Example of this program is shown below. This program performs the FFT on your time series data (stored in the red bin) to produce the frequency domain data, which is stored in the green bin in this example program.

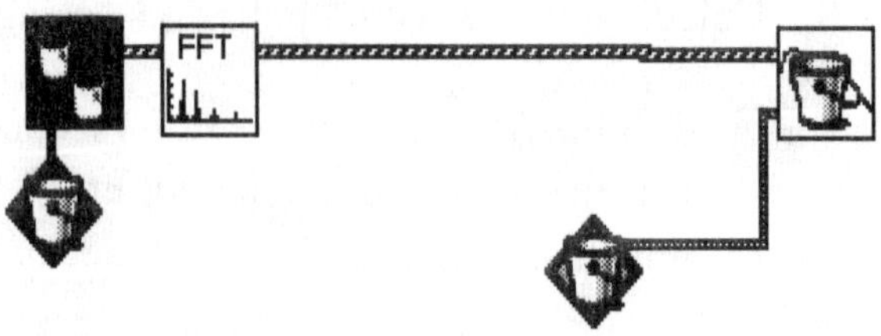

Figure 4.6.5: Program written in Compute Tools to perform the Fast Fourier Transform (FFT) on the data set.

When the result of the FFT is plotted, the natural frequency is evident as the most dominant peak in the graph. (See Extension 4.5.9)

4.7 Terminal Velocity

Age group: 14+ years

Essential Question:

Why does a skydiver reach constant velocity, while the force of gravity continues?

Recommended Time: 40-50 minutes

Materials:

NXT or RCX kits (parts for a single motor car)
Mindstorms or Robolab software
LEDs (Light Emitting Diodes used as a load resistor)
Rotation sensor/ encoder
Ramp with adjustable height

Activity Overview:

The acceleration of any object is directly proportional to the applied Net Force.
In free fall in the air, the object's acceleration is close to the acceleration due to gravity ($g = 9.8\ ms^{-2}$) if the air resistance can be neglected. However, as we know, when the object's speed is increasing, the force of air resistance increases as well, so the net force is decreasing. This is why a skydiver's velocity increases initially and then reaches a constant value, called terminal velocity. The velocity is not changing anymore, so the acceleration is zero, and the net force is zero too.

In this experiment the situation will be modeled using a LEGO car running an electrical LEGO generator (LEGO motor) with a load resistance connected to its contacts. The force of resistance of the motor increases with increased speed. This results in the car achieving terminal velocity.

To create a constant acceleration similar to freefall, an incline or ramp will be used.
To vary the acceleration and compare terminal velocities, the angle of the ramp can be changed.

Learning Objectives:

Students will learn from this experiment that for an object to reach terminal velocity, the resistant force changes and increases with the speed of the moving object.

Building Ideas:

1. Build a simple one-motor LEGO or NXT car.
2. Connect LEDs to the contacts of the motor. (See Figure 4.7.1)

Figure 4.7.1 RCX Model car and LEDs attached to motor
Note: the motor is **not connected** to the RCX, but running as a generator through the interconnected gears of motor and wheel.

Activity Instructions:

1. Program your car to collect rotation sensor data.
2. To set up an inclined plane; take any 2-3 ft. long board and install it at an angle of 10-30 degrees.
3. Before you start the experiment on the incline, determine which direction of motion of the car will cause the LED to emit light due to its polarization.
4. Run the car down the ramp **both ways** so that you can see the change in speed of the car with the motor running and "LEDS on" versus the motor off and "LEDS off". See Figures 4.7.2 and 4.7.3.
5. Compare both graphs.
6. Determine the terminal velocity from the "LEDS On" graph.

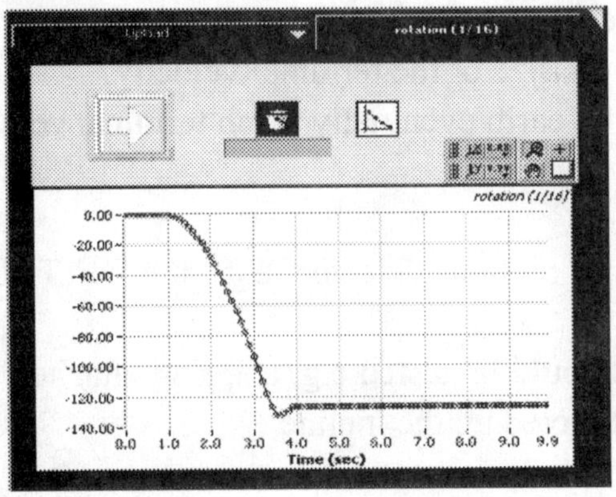

Figure 4.7.2 "LEDS OFF"

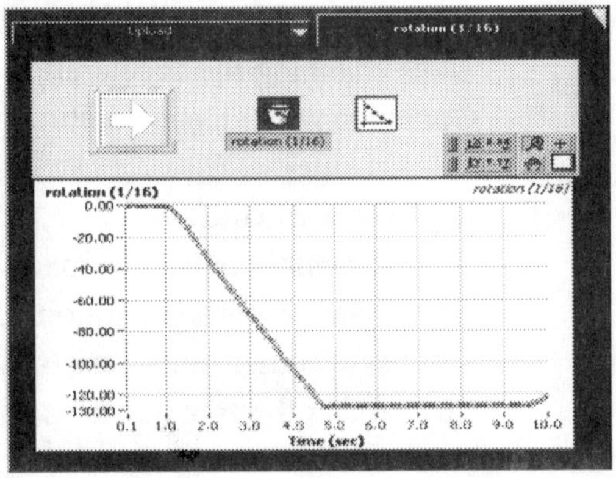

Figure 4.7.3 "LEDS ON"

7. Decrease the angle of the incline and with "LEDS ON", record the data to determine the new Terminal velocity.
8. Increase the angle of the incline and with "LEDS ON", record the data to determine the new Terminal velocity.

Student Observations and Analysis:

Experiment description	Observations	Data Collection
Experiment 1: Car is going down the incline and LED is not connected to the motor	What can you say about the speed of the car??	From the graph of motion what you can conclude about the speed of the car?
Experiment 2: Connect LED and turn on.	What can you say about the speed of the car?	From the graph of motion what you can conclude about the speed of the car?
Experiment 3: Decrease the angle of the incline and repeat Experiment 2	What can you say about the speed of the car?	From the graph of motion what you can conclude about the speed of the car?
Experiment 4: Increase the angle of the incline and repeat Experiment 2	What can you say about the speed of the car?	From the graph of motion what you can conclude about the speed of the car?

Reflection Question:

1. What is the relationship between the resistance force and the speed of the car?
2. Does changing the acceleration affect the size of the terminal velocity?
3. Conclude why all accelerating bodies on earth eventually reach terminal velocity?

Notes and Extensions:

1. Comparisons could be made with a motion detector e.g. the NXT LEGO Ultrasonic or Vernier Motion sensor.
2. Discuss whether other resistant forces could be tested e.g. a spring attached to the cart, so that the force of the spring is directed up the incline.

Chapter 5

Force and Motion

5.1 Gear Ratios and Speed

5.2 Gear Ratios and Force

5.3 Newton's Laws Demonstration

5.4 Newton's 2nd Law Investigation

5.5 Measuring Friction

5.1 Gear Ratios and Speed

Age group: 14+

Essential Question:
Why do we need to change gears?

Recommended Time: 40 – 50 minutes

Materials:

LEGO Gears
2 RCX rotation sensors
RCX or NXT.
1 LEGO motor or 2 NXT Motor /encoders

Activity Overview:

The challenge of this activity is to investigate gear trains and relate them to speed. From this and the next activity called Gear Ratios and Force (*see* Activity 5.2), you can design an efficient Crane gantry (*see* Activity 7.6).

Data logging an angle sensor attached to the axles can determine total angle a wheel moves in a certain time. With two angle sensors, you can measure the motor axle angle versus the wheel axle angle. Because for all machines:

$$\text{Speed ratio} = \frac{\text{Distance moved by effort}}{\text{Distance moved by load}}$$

In terms of a gear train:

$$\text{Speed ratio} = \frac{\text{Angular distance turned by motor gear}}{\text{Angular distance turned by wheel gear}}$$

Learning objectives:

1. Analysis of situations with a number of forces.
2. Model and analyze energy transformation.
3. Design and perform controlled investigations.
4. Refine investigations after evaluating variations and inconsistencies.
5. Analysis of applications of simple machines.
6. Use mathematical ratios to verify physical laws.

Building Ideas:

1. Set up a gear train to test various gear ratios (*see* Figures 5.1.1 and 5.1.2).
2. Put the rotation sensors/encoders at various positions along the train.
3. The rotation sensor on the motor axle will measure motor speed.
4. The rotation sensor on the end gear will measure wheel speed (no need to attach a wheel).
5. For RCX: setup the motor opposite one rotation sensor to drive the system.
6. For NXT: use one of the motor/encoders to drive the system.

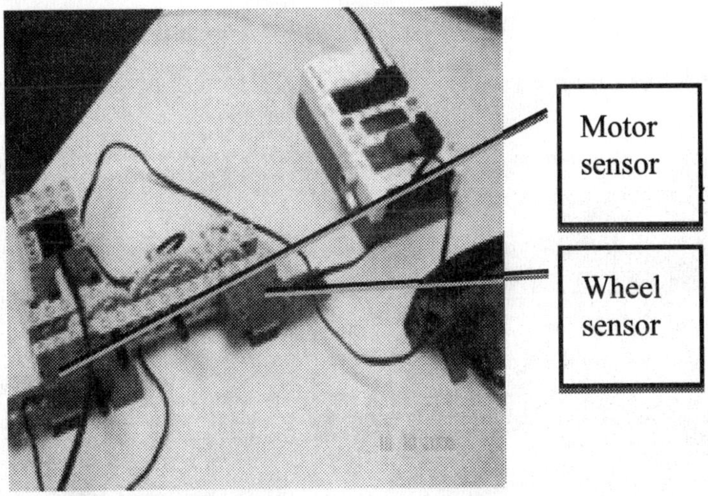

Figure 5.1.1: RCX set-up

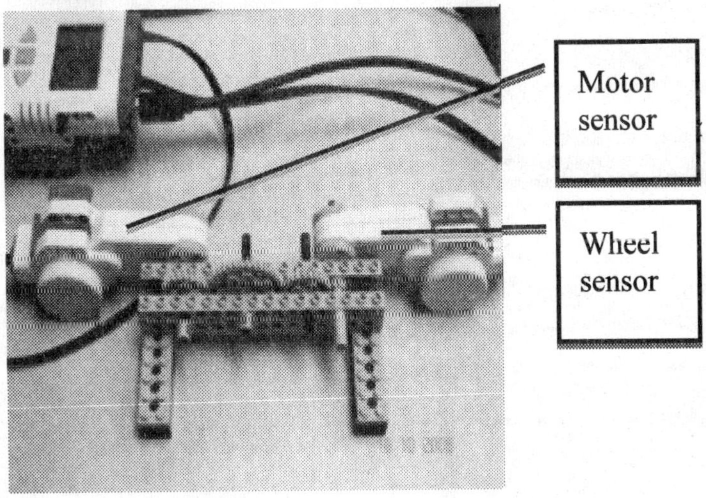

Figure 5.1.2: NXT set-up

Activity Instructions:

Once your gear train is built and set up with sensors and motor to go, attach these to the correct ports on the RCX (1 & 3 rotation, A for Motor) or NXT (A and C for motor encoders). **Record the number of teeth for each gear along the train.**

Programming Ideas
ROBOLAB Investigator

1. Use program level 4 or 5.
2. Zero and data log the two encoders/ rotation sensors (*see* Figure 5.1.3).
3. Program the motor to run for a given time.
4. Download and run.
5. Upload data sets.
6. Using Compute tools determine the ratios of motor angle/ wheel angle (*see* Figure 5.1.4).

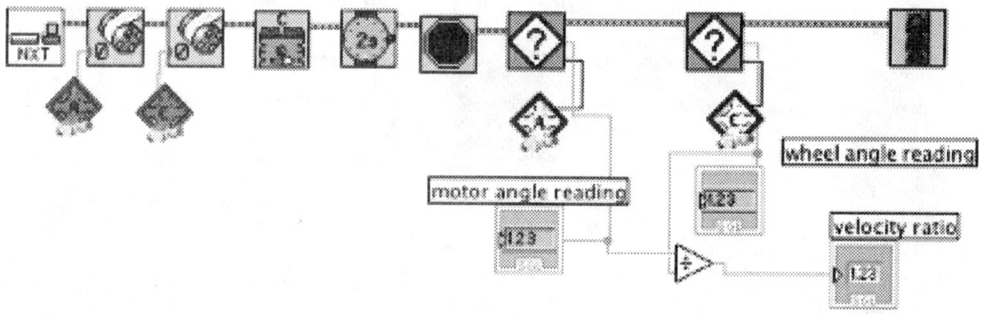

Figure 5.1.3: Robolab Program

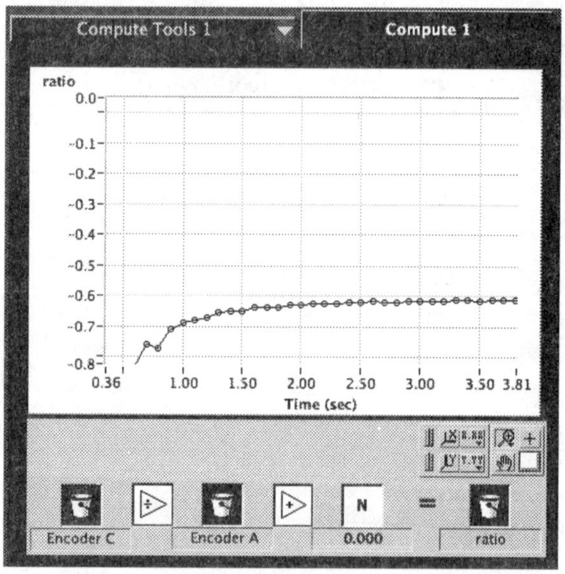

Figure 5.1.4: Robolab data for speed ratio

ROBOLAB Front Panel Instructions:

1. You can run a program available at our website, http://www.physicswithrobotics.com -- *gear ratios NXT.vi or gear ratios RCX.VI*
2. Go to the **Front panel** to run the program (*see* Figure 5.1.5).

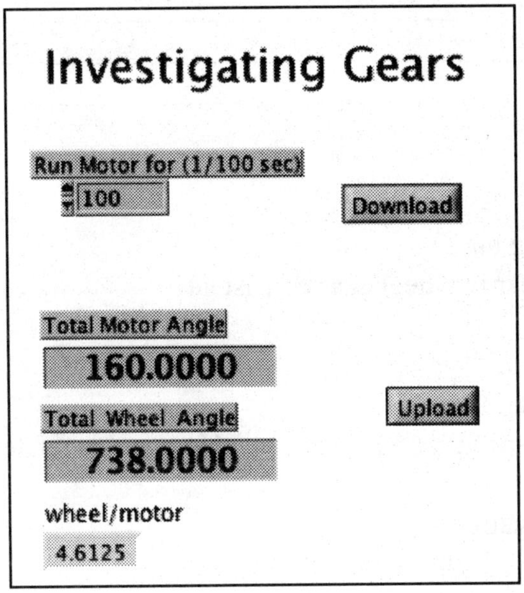

Figure 5.1.5: ROBOLAB Front Panel Screen

3. You must input a time and download when ready.
4. For RCX: Check the IR tower is seeing the RCX continuously.
5. For NXT: the USB link remains attached.
6. Once the gear train has run, there will be non zero outputs from each sensor and the wheel/ motor ratio.
7. If there is a zero reading it may be:
    a. Your connections are incorrect (wrong ports or loose wires)
    b. The sensor is running in reverse (negative values and this program gives the maximum 0) – reverse the sensor.
8. Run a number of times on the same two gears. Try changing the time length.
9. Record the average speed ratio from the computer for that particular wheel gear. E.g. if the sensors are on the first and last gears record this on gear #4.
10. Now move the wheel sensor to a different gear along the train and repeat the experiment.
11. Continue until all gears are tested.

Student Observations and Analysis:

Gear #	No of teeth	Speed = motor angle / wheel angle
1 (Drive gear)		1
2		
3		
4		

Table 5.1.1: Data Table

Reflection questions:

1. Which turns faster small gears or big ones?
2. Is the speed ratio of drive (motor) gear to wheel gear consistent?
3. Does length of time affect it?
4. Is the speed ratio related to teeth ratio?
5. Do the in between gears affect this ratio?
6. For any given gear train, how is the speed ratio determined? How is this related to teeth?
7. What did you learn from this investigation?

Notes and Extensions:

1. Does the type or number of gears between the motor and wheel gears have any affect on speed? To verify this mathematically check the following (*see* Table 5.1.2):

Example

	Drive gear	Gear 2	Gear 3	Wheel gear
No of teeth	8	24	40	24
Multiplying ratios		8/24 = 1/3	1/3 × 24/40 = 1/5	1/5 × 40/24 = 8/24

Table 5.1.2: Mathematical analysis of gear ratios

Now try your combination and complete Table 5.1.3

	Drive gear	Gear 2	Gear 3	Wheel gear
No of teeth				
Multiplying ratios				

Table 5.1.3

2. These investigations are valuable for teachers wanting students to understand the simple machines and speed as well discovering some interesting facts about gear ratios.

3. Note that something can't be gained without something else being reduced and in this case its speed versus force. In Chapter 5 Gear ratios and Force is investigated.

4. This would also be great for lessons in math needing some practical applications of ratios.

5. This activity along with your work on Gears and rotational force in Chapter 5 is preparation for the Engineering Challenge in Chapter 7: Design a Motorised Crane Challenge.

5.2 Gear Ratios and Force

Age group: 14+

Essential Question

How do gear combinations reduce force?

Recommended Time: 60-90 minutes

Materials

LEGO materials
2 pulleys and string
50g Weights
Force sensor or spring balance.

Activity Overview

This activity involves a simple construction of gear trains with weights or balances applied at each end to determine the mechanical advantage (MA) and relate it to the gear ratio.

Learning Objectives

1. Analyse situations with a number of forces.
2. Model and analyze energy transformation.
3. Design and perform controlled investigations.
4. Refine investigations after evaluating variations and inconsistencies.
5. Analyze applications of simple machines.
6. Use mathematical ratios to verify physics laws.

Building Ideas:

1. Set up a gear train to test various gear ratios (*see* Figure 5.2.1).
2. Put pulleys with thread on each end.
3. You can hang a force sensor or spring balance at one end and various weights at the end of the train or use weights on both sides!

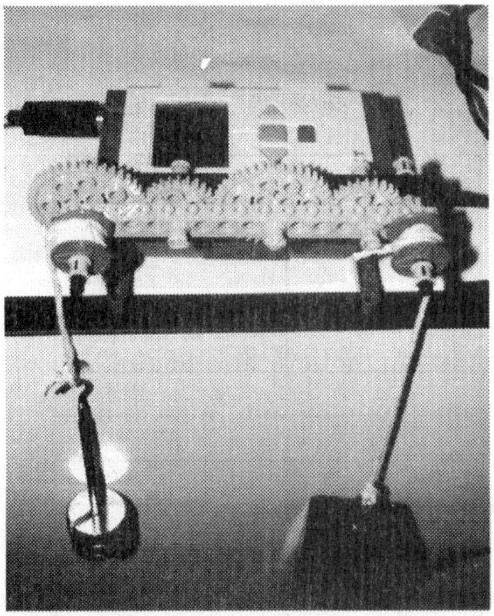

Figure 5.2.1: Experiment set-up.

Activity instructions

1. When your gear train is built and set up with pulleys, cord and hooks, attach 100g weight on load gear (large) and a force sensor or balance on effort gear (small). (*See* Figure 5.2.2).

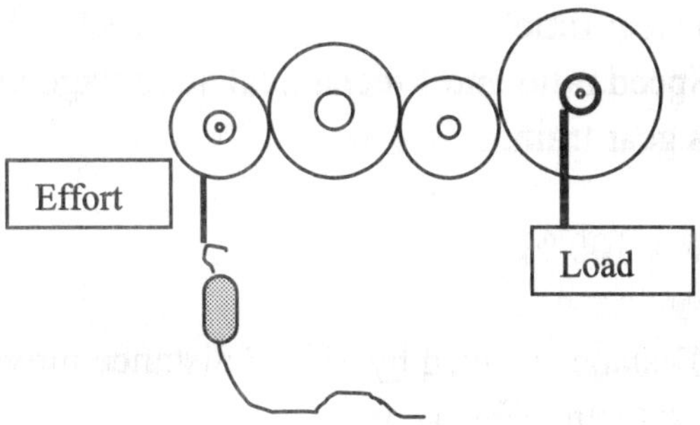

Figure 5.2.2: Gear train example

2. Record the force required to just move the weight upwards.
3. Add another 100g and record force.
4. Add another 100g to record three results.
5. Now try the next gear along the train for the effort and repeat the above 3 trials.
6. Try swapping the load and effort gears.

Student Observations and Analysis

Gear ratio (# teeth load / # teeth effort)	Weight added	Force required	MA (load/effort)

Table 5.2.1: Experimental data

8. Is the Mechanical Advantage ratio consistent for each gear ratio?
9. Does the increase in weight added affect the efficiency?
10. Is MA related to Gear ratio?
11. Combining the Speed ratio and Mechanical advantage you can determine the efficiency of this gear train:

Efficiency = MA /SR x 100 %

Speed ratio (SR) = Distance moved by effort/ distance moved by load
= # teeth load gear/ # teeth effort gear

Determine the efficiency of your gear train!

Reflection Questions

1. To reduce effort required and so use a ratio of gears with the greatest mechanical advantage what is the best combination?
2. How efficient is the system?
3. What can you do to increase efficiency?

Notes and Extension

1. In the Activity in Chapter 4: Gear Ratios and speed, the speed ratio increased with gearing up the number of teeth on the motor cog. There is a compromise however, as students will often discover in attempting to increase the speed of their super dragster. Gaining speed loses rotational force or torque.
   Using Moments about the centre of a gear and transferring force to a meshed gear, it can be shown:

   $$\frac{F_1}{F_2} = \frac{r_1}{r_2}$$

   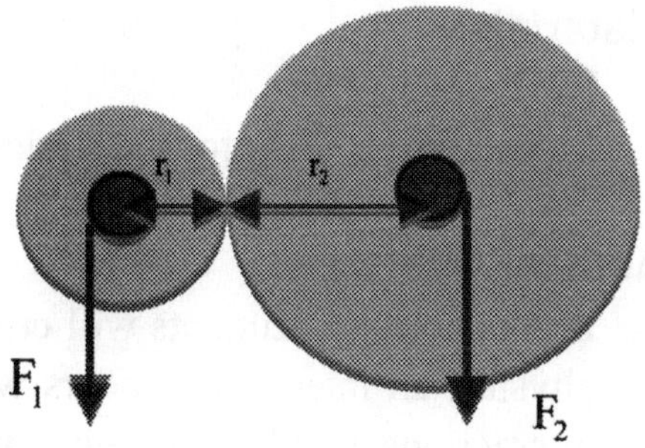

   This means the smaller the gear on the wheel and greater speed ratio, the weaker the force!
2. Mechanical Advantage is also an important concept for machines to show the effort force can be reduced.
   MA = Load / Effort *or for gears:*
   MA = Output force from Wheel/ Input force from Motor.
   MA > 1 for a geared down system.
3. We've tested speed ratio and noted that it is only dependant on teeth number. However, the force ratio is affected by internal forces of friction and tension in the string; hence, efficiency should also be determined.
   Efficiency = MA/ speed ratio × 100 %
4. A mechanical hoist or crane can now be constructed. See Chapter 7: Design a Motorised Crane Challenge.
5. For advanced classes, it is worthwhile to calculate Energy efficiency based on Input Power and Output power. For a motorised Crane:
   Power Input = Voltage supplied × Current Drawn.
   Power Output =Work done/ time = $\frac{\text{mass lifted} \times g \times \text{height raised}}{\text{Time required}}$.
   Energy Efficiency = PO/PI.
   Voltage and current can be measured using multimeters of Vernier sensors.

5.3 Newton's Laws Demonstrations

Age Group: 14+

Essential Questions: What can a robot teach you about Newton's three laws of motion?

Recommended Time: 2-3 class periods

Materials:

NXT or RCX kit
Measurement tools for the students to support their demonstrations with data

Activity Overview:

In this activity, students will create a robot system that interacts with the real world. In physics, an interaction occurs when objects exert a force on one another. Examples of interactions could be pulling a wagon, pushing on a wall, or sliding across the floor. It is not hard to create an interaction because any time an object's motion changes, it is the result of an interaction. In this activity, it is up to the student to come up with creative interactions that helps them demonstrate the meaning of Newton's three laws. One important aspect of this activity is that students support their demonstrations with data.

Learning Objectives:

By the end of this activity, students will be able to:

1. Demonstrate interactions between a robot system and objects in its environment.
2. Demonstrate the application of Newton's three laws to a robot interaction.
3. Use data to support the demonstration of Newton's three laws.

Pre-Activity Discussion Questions:

1. In physics, what are interactions?
2. What are examples of interactions in everyday life?
3. Pick a few interactions and describe how all three of Newton's three laws apply to these interactions?

Building Ideas:

The details of the systems the students will build depend upon their ideas for demonstrating the laws. If the students have limited building experience with the LEGO systems, we recommend they start with either the single motor or double motor systems shown in Chapter 2 of this book.

Programming Ideas:

While the details for the programming of the robot systems built by your students will depend upon their specific ideas, it is likely that the programming will consist primarily of turning on a motor at a certain power level for a certain amount of time. One useful programming technique for this activity is to have the robots "coast" to a stop rather than stop abruptly. Figures 5.4.1 and 5.4.2 show code in the Mindstorms and Robolab software environments.

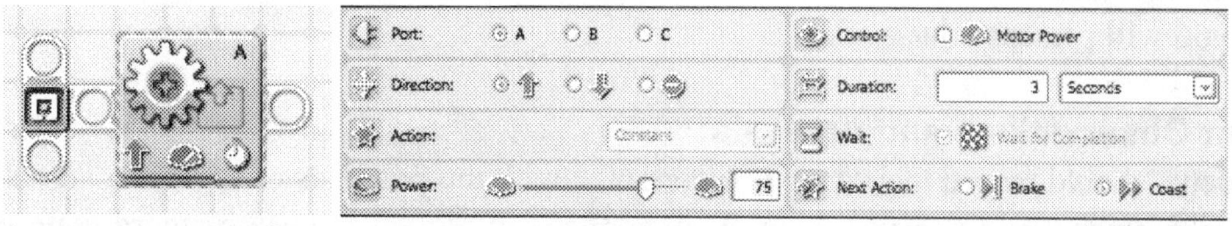

Figure 5.4.1: Mindstorms code to turn on motor A at 75% power for three seconds and then let it coast to a stop.

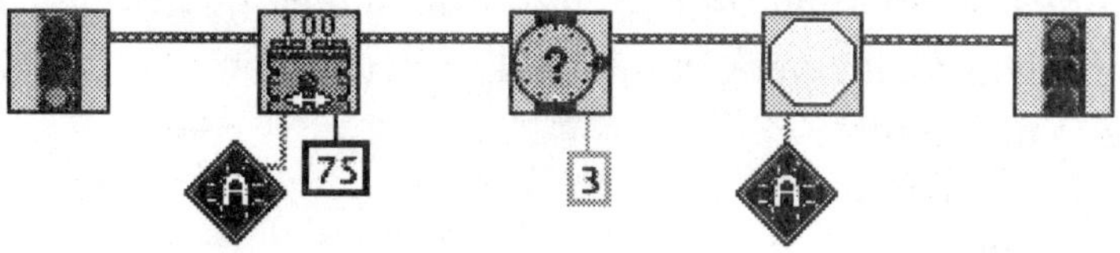

Figure 5.4.2: Robolab code to turn on motor A at 75% power for three seconds and then let it coast to a stop.

Activity Instructions:

1. This activity can be done in three different ways. You can:
    a. Apply all three laws to a single interaction – a robot skidding to a stop, for example. How can you use this single interaction to demonstrate all three of Newton's laws?
    b. Apply each law to a different interaction – the robot starting to move for Newton's 2nd law, the robot moving at a constant velocity for Newton's 1st law, and the robot stopping for Newton's 3rd law.
    c. Conduct this activity with a combination of *a.* and *b.* above.
2. Review Newton's three laws and sketch your ideas for building a LEGO system that demonstrates all three laws. If you have done other projects in this book, think about how the laws apply to those projects. If this is your first project, think about how a LEGO device can exert forces on other objects (LEGO or non-LEGO) and how those objects exert forces on the LEGO system.
3. Brainstorm ideas for how you will collect data to support your demonstrations. What will you measure and how? If you do not have the ability to collect data to support your ideas, return to step one and brainstorm a new interaction. NOTE: Do not underestimate the time required by steps 2 and 3. Plan your efforts well to ensure high quality interactions supported by high quality data.

4. We recommend that you take the time to share your ideas with the class. It is likely that most groups of students will be conducting different demonstrations. However, sharing ideas related to data collection may be very useful. Data collection methods for supporting a demonstration of Newton's 3rd law may be identical even if the actual interactions are very different.
5. Once you are ready, build your system that demonstrates Newton's laws. Collect data to support your demonstrations. Consult with your instructor to determine how you will present your work.

Student Observations and Analysis:

Students should record their data in accordance with the practices of the class. We do not recommend any specific system because the method you choose to record your measurements depend highly on the system you choose for the actual data collection. A student is likely to record data differently from a spring scale force meter compared to a student who collects data from a force probe connected to a computer.

Keep the final presentation of the student work in mind when deciding on the best way to record and represent data in this activity.

Reflection Questions:

1. How will you and your classmates present your demonstrations and data?
2. Which law did you find the easiest to demonstrate? Describe what you did and how you used data to support the demonstration.
3. Which law did you find the hardest to demonstrate? Describe what you did, why it was hard. Describe how you used data to support the demonstration.

Notes and Extensions:

1. Students can present their demonstrations in many different ways. Some examples are:
    a. A traditional lab report
    b. A classroom demonstration of the interactions and presentation of data that supports the application of each law.
    c. An audiovisual presentation – a digital movie, or slideshow that stands alone to communicate how the student's projects demonstrated Newton's laws and how these demonstrations were supported by data.
2. An alternative approach to using traditional forms of data (i.e. numerical data) is to use qualitative observations as evidence for each law. For example, attaching a LEGO vehicle to a small wagon can demonstrate Newton's first law. Place a marble in the wagon. When the LEGO vehicle moves forward, the marble remains at rest. Relative to the wagon, it moves *backwards.* When the wagon stops, the marble stays in motion until it strikes the front of the wagon.

5.4 Newton's 2nd Law Investigation

Age group: 16+ years

Essential Question:

How can we change the velocity of a moving object?

Recommended Time: 40 - 50 minutes

Materials:

NXT or RCX Kit with rotation sensor
Set of weights
Pulley

Activity Overview:

This activity will introduce students to the relationship between the acceleration, force, and mass. Students will experiment with the motion of an object when the force is applied to it.

Students will qualitatively and quantitatively find that the acceleration of an object increases proportionally to the net force acting on the object and inversely proportionally to the mass of the accelerating system.

$$a = \frac{F}{m}$$

$$F = ma$$

Learning Objective:

Develop an understanding of the relation between acceleration, force, and mass.

Building Ideas:

1. In this experiment you will use NXT/RCX car, which will be accelerated by the external force applied to it.
2. You need also a pulley, a string, and a hanger with a set of masses to vary the forces applied to the car.
3. You need rotational sensor for this experiment. For the NXT, the rotational sensor (encoder) is part of the motor. For the RCX, do not install a motor, but attach a rotational sensor to one of the axes instead.
4. Experimental setup is shown in Figure 5.3.1.

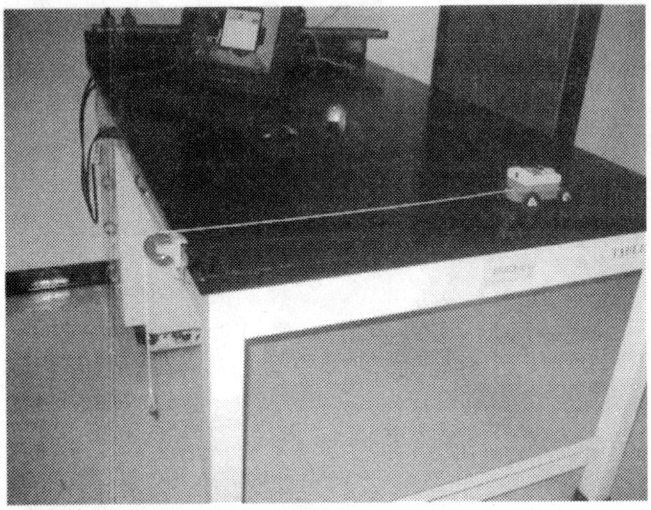

Figure 5.3.1: Experiment arrangement

Activity Instructions:

1. Program the NXT/RCX to collect the rotation sensor/encoder readings.
2. Measure the mass of the NXT/RCX car, M_1______________
3. Measure the mass of the hanger with additional masses, M_2____________
4. In your experiment you will distribute additional masses between the hanger and the NXT/RCX car. As a result the total mass of moving system will be kept constant ($M = M_1 + M_2$), but the force applied to the system will be equal to the weight of the hanger with remaining masses on it.
5. Transfer some of the masses from the hanger to the NXT/RCX car and record the mass of the hanger with remaining masses on it (m) in Table 1.
6. Start collecting data from the NXT/RCX and release the system.
7. Upload the data from the encoder (NXT) or rotation sensor (RCX).
8. Repeat steps 5 and 6 for different distributions of masses between the NXT/RCX and the hanger.
9. See Figure 5.3.2 for the sample graph from the rotation sensor data.

Student Observations and Analysis:

Table 1

	Total hanging mass (m)	Total mass of the system ($M=M_1+M_2$)	$F_{net} = mg$ (neglecting friction)	$a_{calculated}$ $a = \frac{mg}{M}$	$a_{experimental}$
Experiment 1					
Experiment 2					
Experiment 3					

Calculate the experimental value of linear acceleration form the angle sensor data. Figure 5.3.2 shows experimental data from the RCX rotation sensor. Note: For the RCX rotation sensor the data is represented in units of $\frac{1}{16}$ of a revolution. For the NXT encoder, the data is represented in degrees.

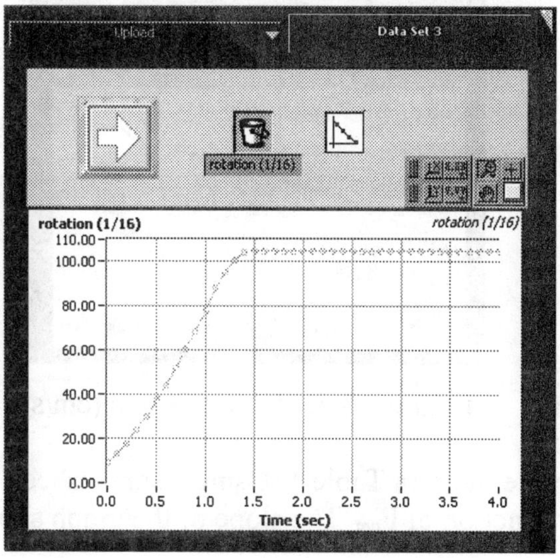

Figure 5.3.2: RCX rotation sensor data.

Convert the angular data into the linear displacement data using:

Linear distance = (number of revolutions) × 2πR

where R is the radius of the wheel of the car

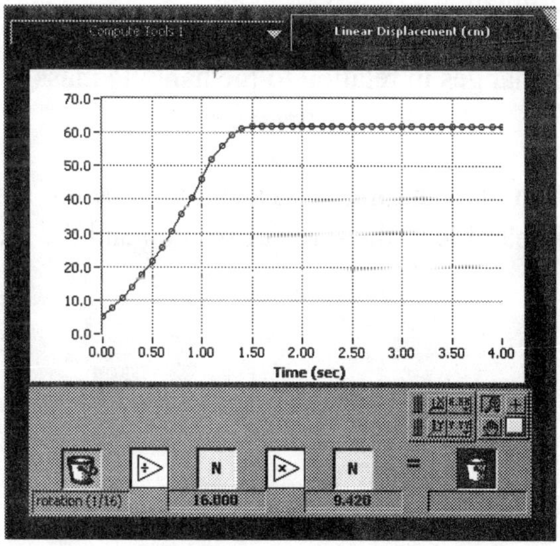

Figure 5.3.3: Graph of Linear Displacement (cm) (R wheel = 1.5cm)

Velocity can be found as the first derivative of the Linear Displacement (*see* Figure 5.3.4). Note that the velocity linearly increases with time. The acceleration can be found as the slope of the velocity graph (*see* Figure 5.3.5).

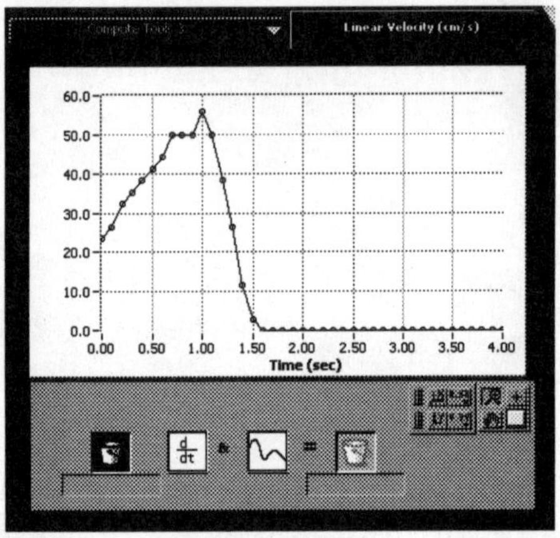

Figure 5.3.4: Graph of Linear Velocity (cm/s)

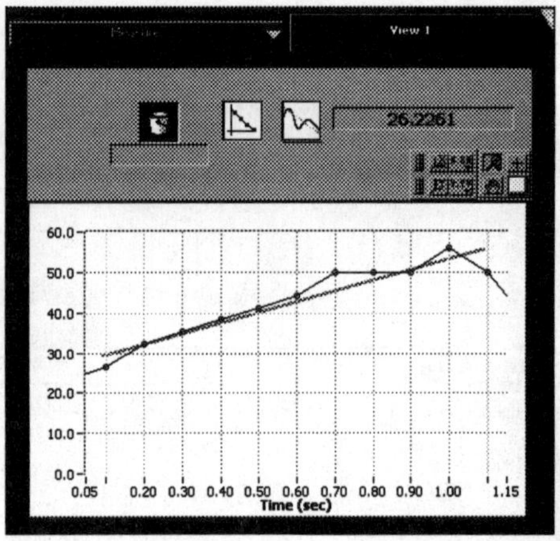

Figure 5.3.5: Value of slope (cm/s^2)

Record your experimental values for acceleration in Table 1. Using a spreadsheet, plot a graph of experimental acceleration as a function of F_{net}. The slope of the graph should be equal to $\frac{1}{m}$, where m is a total mass of the system.

Reflection Questions:

1. How were you able to control the force applied to the system?
2. Did the vehicle in your experiment have a constant speed or not? Please explain your answer using the data from your experiment.
3. If the total mass of the system remains the same (by transferring weights from the car to the hanger), how does acceleration changes in relation to the hanging mass?

Notes and Extensions:

In order to decrease the effect of the friction, the hanging mass should not be too small. Otherwise, the pulling force would be comparable to the force of friction and the motion is more complicated for analysis.

5.5 Measuring Friction

Age Group: 16+

Essential Questions: Can a robot help me learn the meaning of Greek symbols?

Recommended Time: 45-90 minutes

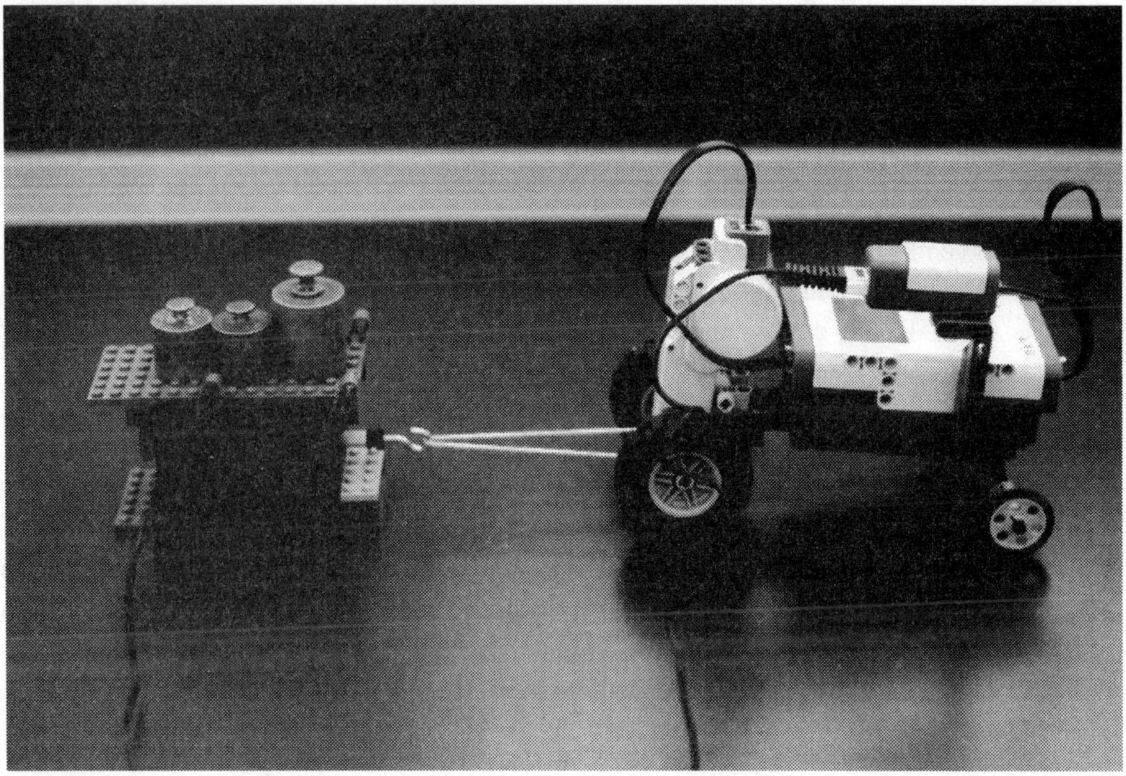

Figure 5.5.1: Experimental setup.

Materials:

NXT or RCX kit
Set of weights
Force probe
Mass or weight scale

Activity Overview:

In this activity, students will use Coulomb's law of friction ($F_f = \mu F_N$) to determine the coefficient of static and kinetic friction for several surfaces such as tabletops, different types of paper, and different floor surfaces, wood. The LEGO robot and a force sensor is an ideal tool for conducting this investigation because it allows for hands free, repeatable experiments.

This activity is suitable as an investigation of Coulomb's law of friction alone or as a supporting activity for an engineering challenge such as a hill climb. See the notes and extensions section of this activity for ideas.

Pre-Activity Discussion Questions:

1. What is friction and what would the world be like without friction?
2. Why is friction important?
3. What factors affect the amount of friction and what equation relates those factors together?
4. When studying the factors that affect friction, we refer to something called the "Normal" force. What is it and how do you measure or calculate its amount?

Learning Objectives:

By the end of this activity, students will be able to:

1. Differentiate and measure frictional forces and normal forces
2. Calculate coefficients of friction
3. Relate Coulomb's law to objects in everyday life

LEGO Sample Building Instructions:

See Chapter 2 for an example of a single motor vehicle that can pull a force meter sled (Figure 5.5.1). To conduct this experiment, the robot only needs to drive straight for a specified amount of time. If you are conducting this investigation with an RCX, we recommend that you build a model with a gear ratio that reduces the speed of the unit while increasing its pulling force. This will help prevent your wheels from spinning and allow you to pull the sled at a constant speed. *See* Chapter 2 for an example RCX car built with gears.

Figure 5.5.2: Parts for the experiment

Figure 5.5.3: Vernier Force Sensors

Sample Programming and Data logging Instructions:

As with most activities in this book, this activity can be completed with different combinations of robot hardware, programming software, and data logging software. If you are using the Vernier Dual-Range Force Sensor, the table below represents your hardware and software options. If you are conducting this activity with another digital force sensor, consult the manufacturer for your options. In general, if your sensor does not communicate directly with the robot hardware then you need to do the programming and data logging separately.

Hardware	Programming	Data Logging
NXT	MINDSTORMS 1.0 – 2.0, ROBOLAB	MINDSTORMS 2.0, ROBOLAB Investigator Vernier Logger *Pro*
RCX	ROBOLAB	Vernier Logger *Pro*

Table 5.5.1: Hardware and Software Options for the Vernier Dual Range Force Sensor

Programming and Data Logging with Mindstorms 2.0

With the release of version 2.0 of the Mindstorms Education NXT Software, the user can program the robot to move and collect data. The following is a sample program used in a friction investigation activity[1] (Figure 5.5.4). Figure 5.5.5 shows example force data from a single motorcar pulling the force meter sled from Figure 5.5.3. Figure 5.5.6 shows an example ROBOLAB program and graph. It is more complicated to analyse the data in ROBOLAB. We do not show the steps to follow here. Please see our website, www.physicswithrobotics.com for more information. If you are measuring force with a separate data logging program like Vernier's Logger *Pro*, refer to chapter 3 for more details.

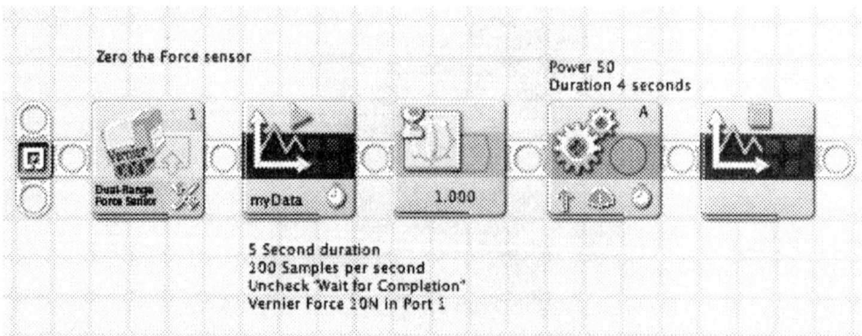

Figure 5.5.4: Mindstorms 2.0 program to collect data while running a motor

[1] If you have not done so already, download the "Vernier Sensor" block from Hhttp://www.vernier.com/nxtH. Install it according to the directions on the website.

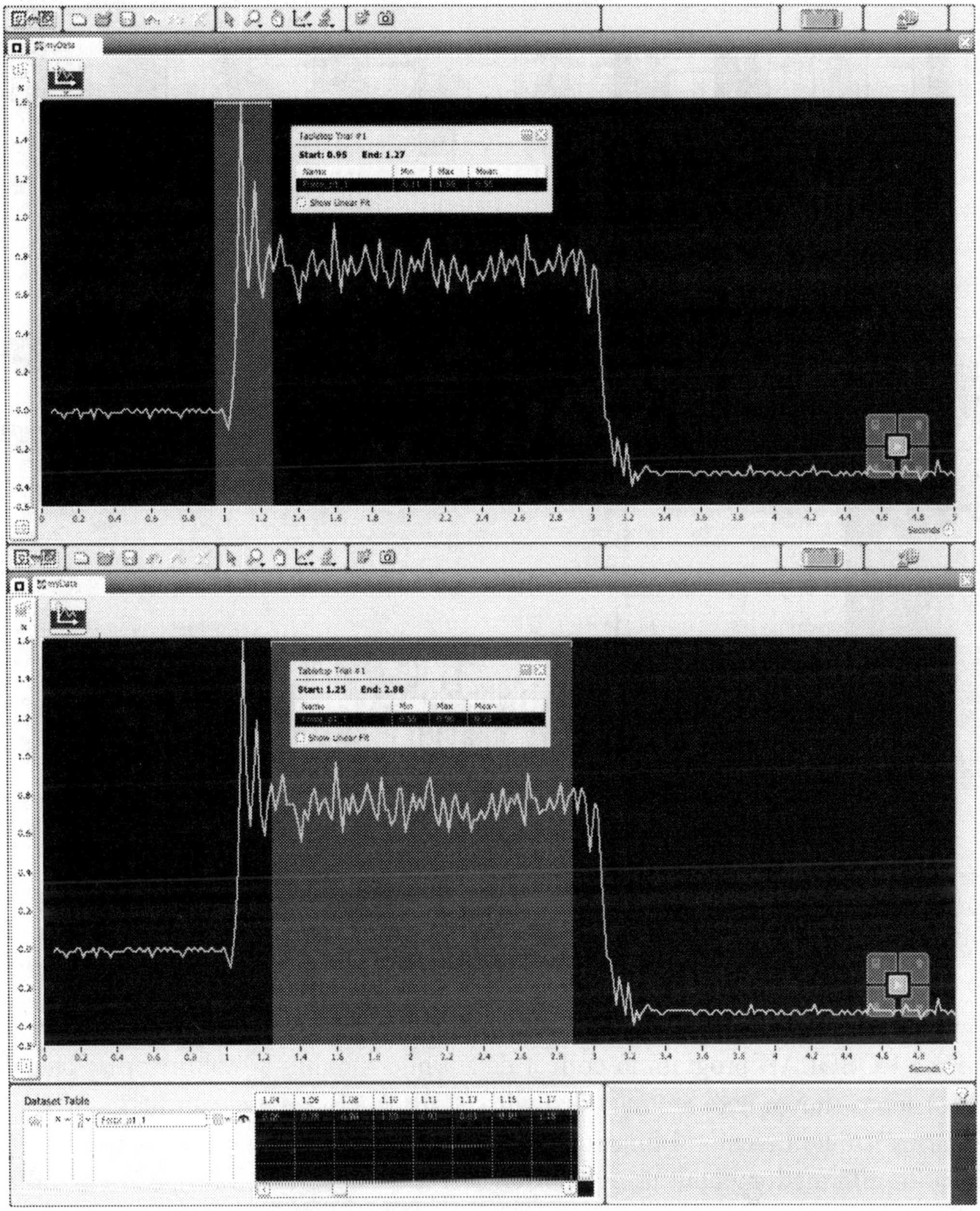

Figure 5.5.5: MINDSTORMS software (v2.0) graph and data table of a single motorcar pulling a Vernier force meter sled. Note the two analysis areas. In the top graph, the peak region is highlighted and analysed for a maximum value, the maximum force of static friction. In the second graph, the equilibrium portion of the graph is analysed to determine the average equilibrium force, which is the force of kinetic friction.

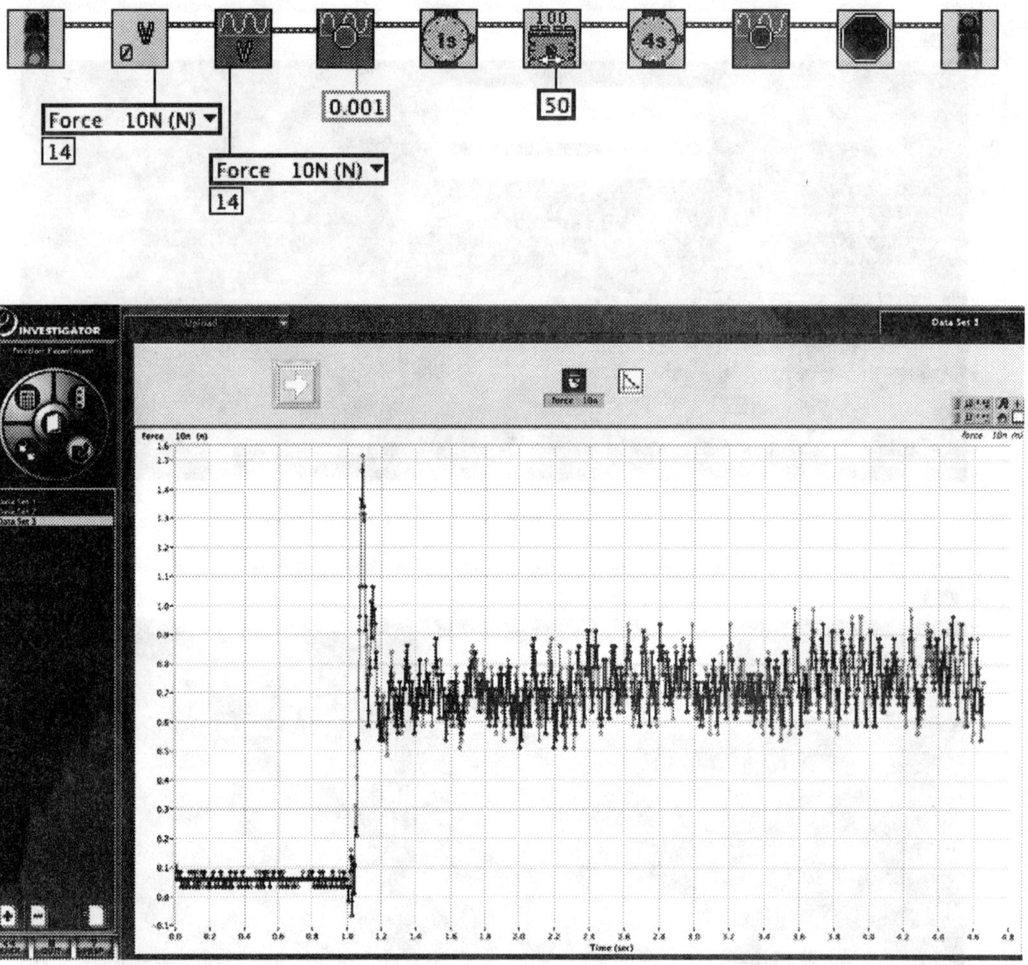

Figure 5.5.6: Data from a Vernier force meter sled being pulled by a single motor car and the ROBOLAB program to collect data while running a motor (top). The Vernier sensor icons shown here are included with ROBOLAB 2.9 and can be found the “User Libraries” of the Inventor 4 functions palette. See chapter 3 and our companion website for more information about using ROBOLAB to analyse data for maximum values and average values.

Activity Instructions:

1. Build a sled that can hold the force meter (see Figure 5.5.3).
2. Configure your robot and data collection software to collect force data while pulling the force sled. Pick your surfaces to test and tune the speed of your robot so that its wheels do not spin while pulling.
3. Determine the time to pull the sled based on your lab surface.
4. Once your robot is pulling without spinning its wheels and pulling at a constant speed, you are ready to collect data.
5. Run your robot while collecting data and observe the resulting graph.
6. Analyse your data to determine the maximum force of static friction and then the average force of kinetic friction. See the Programming and Data logging section of this activity for examples. Record the results of your analysis in the Student Observations and Analysis section of this activity.
7. Repeat steps five and six 3-5 times and obtain an average for both the force of static friction and kinetic friction.
8. Measure the mass of your force sled and determine the normal force (F_N) acting on the sled. Record the results of your analysis in the Student Observations and Analysis section of this activity.
9. Use Coulomb's Law ($F_f = \mu F_N$) to determine the coefficients of static and kinetic friction.
10. Repeat these steps for all surfaces being tested.
11. See the Notes and extensions of this activity for application and design challenges connected to this friction analysis.

Student Observations and Analysis:

Data Table 1

Surface	**Mass of the Sled**	F_N	$F_{fstatic}$ Trial 1	$F_{fstatic}$ Trial 2	$F_{fstatic}$ Trial 3	μ_s

Data Table 2

Surface	**Mass of the Sled**	F_N	$F_{fkinetic}$ Trial 1	$F_{fkinetic}$ Trial 2	$F_{fkinetic}$ Trial 3	μ_k

Reflection Questions:

1. In everyday life, where is it important to use surfaces with high coefficients of friction?
2. In everyday life, where is it important to use surfaces with a low coefficient of friction?
3. Did any of the results of this investigation surprise you? If so, explain.
4. Does the coefficient of friction depend on the mass of the objects sliding over one another? Describe an experiment you could use to investigate this question.
5. At any point in your investigation did your wheels spin, preventing your vehicle from pulling the sled? What did you do to resolve the situation? Describe your solution with the terms used in this lab investigation: *friction, normal force,* and *coefficient of friction.*

Notes and Extensions:

1. One subtle design characteristic of this experiment is the angle of the string between the force probe. If the string is not parallel to the ground, the Normal force on the LEGO vehicle's back wheels will be affected.
2. Friction data for LEGO wheels can be found on Philo's website: http://www.philohome.com/traction/traction.htm
3. Hill Climb. Ask students to design a vehicle that can climb the steepest angle. Knowing the coefficients of static and kinetic friction and using Coulomb's law of friction, students can predict the maximum angle to attempt.
4. Tractor Pull. Ask the students to design a system to pull the most weight across a surface. Be careful with this one as LEGO systems can be geared to produce very large forces. See http://www.philohome.com/hoist/hoist.htm for more information.

Chapter 6

Energy

6.1 Microphone Sound Reduction

Age Group: 14+

Essential Questions: If a tree falls in the forest, how do we quiet the sound?

Recommended Time: 90 minutes

Materials:

NXT	NXT Sound sensor
NXT Motor	9 volt buzzer[1]

Activity Overview:

In this activity, students will explore sound waves. Specifically, they will use the NXT sound sensor to explore how sound waves work by designing experiments to investigate the variables that affect the measurement of a sound wave's loudness. For instance, how does the distance between the sources of a sound affect the measurement of the sound wave's loudness? How does the orientation (direction) of the sound source affect the measurement of loudness? How do obstacles between the source and sensor affect the measurement? Are there other variables that affect the measurement of a sound's loudness? It is likely that students will be able to generate hypotheses that describe relationships between variables; the point of the activity, therefore, is to provide them with practice at designing a controlled experiment that either supports or fails to support their hypotheses.

While designing their experiments, conducting observations, and drawing conclusions, students will explore what loudness is and how it is measured. Before they begin, however, students should review the relevant units of measurement for a sound wave's loudness. As we point out in Chapter 2, the NXT sound sensor does not report sound level readings in true decibels (dB). This does not, however, make the sensor less useful for the investigations in this activity. It does mean, however, that students should be aware of the units used by the NXT sound sensor if they wish to develop specific mathematical relationships between variables[2].

---

[1] The 9-volt buzzer is not required but was used in the design of this activity because it is both loud and easy connect to the NXT and program via a motor command. Radio Shack part # 273-075 (95 dB Panel Piezo Buzzer) or similar products will work well for this activity.

[2] If a more detailed treatment of the NXT sound sensor is needed, consult Claude Baumann's work at http://www.convict.lu/htm/rob/NXT_sound_sensor.htm and Brian Davis' work at http://www.brickshelf.com/cgi-bin/gallery.cgi?i=3348152 .

Pre Activity Discussion Questions:

One of the biggest challenges to students and teachers about the topic of sound is that in spite of its omnipresence, we cannot *see it!* An exceptional visual and interactive model of sound waves can be found online, however. We encourage you to visit the interactive simulation at the web address below as a companion to the discussion questions. http://phet.colorado.edu/simulations/sims.php?sim=Sound

1. What is the difference between a sound wave's wavelength, frequency, and amplitude? Which of these best describes the loudness of a sound?
2. How does sound transfer energy? In other words, how does a sound do work? How do the terms energy, power, intensity, and relative intensity?
3. What direction do sound waves move from a source?
4. Assuming that there is no energy lost by a sound wave, why is it not as loud the further away you move from a source?

Learning Objectives:

By the end of this activity, students will be able to:

1. Describe sound waves in terms of a transfer of energy
2. Measure the loudness of a sound wave and relate this measurement to a wave's amplitude, energy, power, and intensity
3. Create an experiment that investigates which variables affect the measured loudness of a sound wave

Building Ideas:

We outline here a few ideas to consider when designing sound experiments for this activity.

1. To help ensure that your sound measurements are measurements of waves traveling through the air, it is necessary to isolate the sound sensor from vibrations that come directly from the sound source motor and from the tabletop. Figure 6.1.1 shows two techniques for isolating the sound sensor. The first technique is to mount the sound sensor on a separate structure and the second technique is to place all equipment on top of packing foam.
2. If students are exploring the effect of an obstruction (i.e. sound absorption) between the source and sound, make sure they consider how sound waves diffract around the edge of the obstruction. This could lead to another investigation.
3. The motor shown in Figure 6.1.1 works well and is well suited for several investigations. The classroom to motor noise ratio, however, may make it impractical to use. Consider using a loud (~95 db) Piezo Buzzer. This ensures a strong signal to noise ratio for your data collection. It can be easily interfaced with the NXT using the NXT to RCX motor/sensor cable. See Figure 6.1.2 for details.

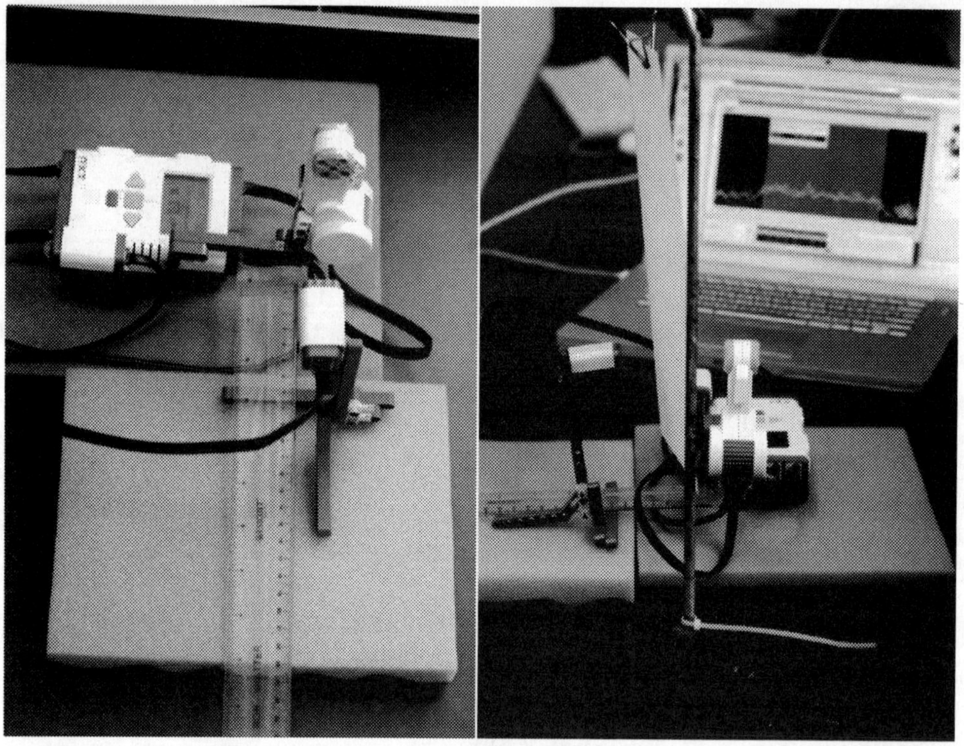

Figure 6.1.1. The design on the left allows for easily varying the distance and relative angle between source (an NXT motor) and sensor. The design on the right shows the addition of an obstruction between the source and the sensor.

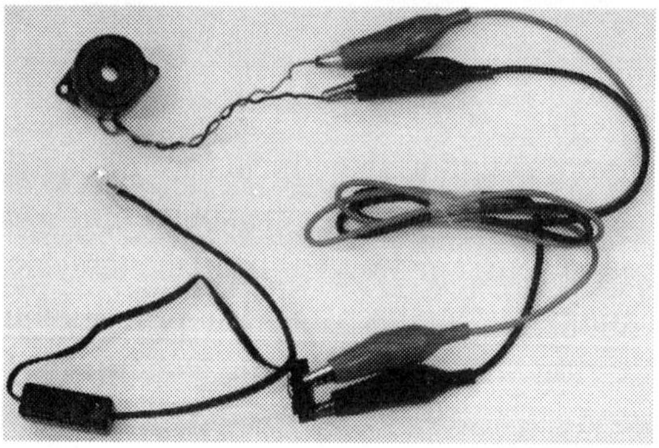

Figure 6.1.2 Wiring for Piezo Buzzer and Legacy Connector

4. Figure 6.1.3 shows an experimental design utilizing the 9V DC Piezo buzzer. You can see in the computer screen that the NXT sound sensor's threshold is reached (left) but when paper "ear" protection is put on the sound levels are reduced considerably. The loudness of the Piezo buzzer allows for flexibility in the design of many experiments. In other words, you do not have to be as concerned about low-level background noise in the classroom.

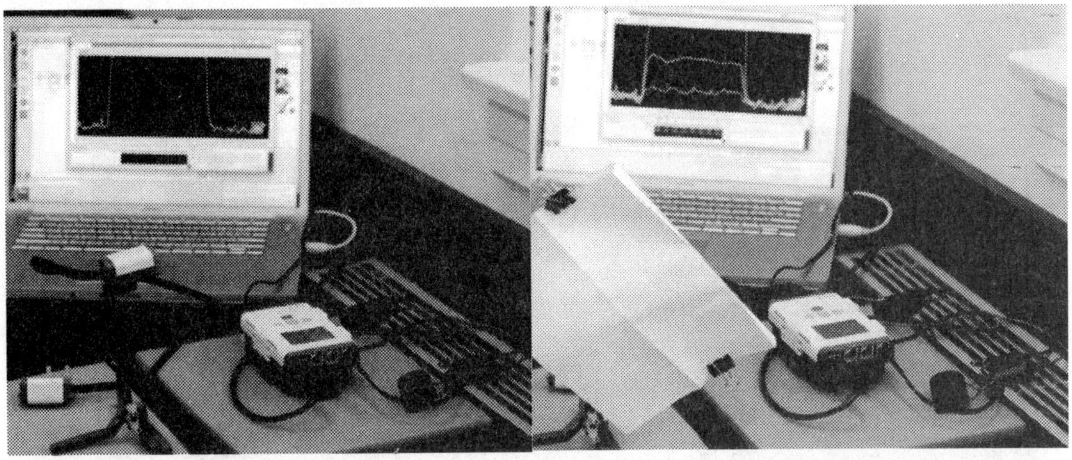

Figure 6.1.3: The Piezo buzzer is used as the sound source here. On the left, we see the results of an experiment with no "ear" protection. On the right, we see the ear protection and the data from two different ear protection designs.

Data Logging Ideas:

Two challenges when collecting sound data with the NXT sound sensor include understanding the units used in measurements and dealing with background noise. To help manage both of these issues, we recommend that students determine baseline conditions and collect baseline data. Figure 6.1.4 shows example Mindstorms 2.0 code for a sound experiment and Figure 6.1.5 represents the data from three baseline runs where the sound

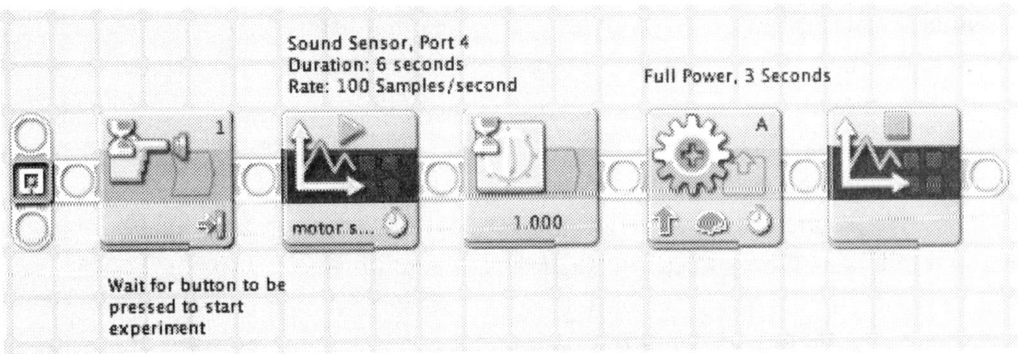

Figure 6.1.4. Mindstorms 2.0 code for collecting sound data from a running motor. This code also works for an experiment using a DC piezo buzzer connected to an NXT motor port via an NXT to RCX motor/sensor converter cable.

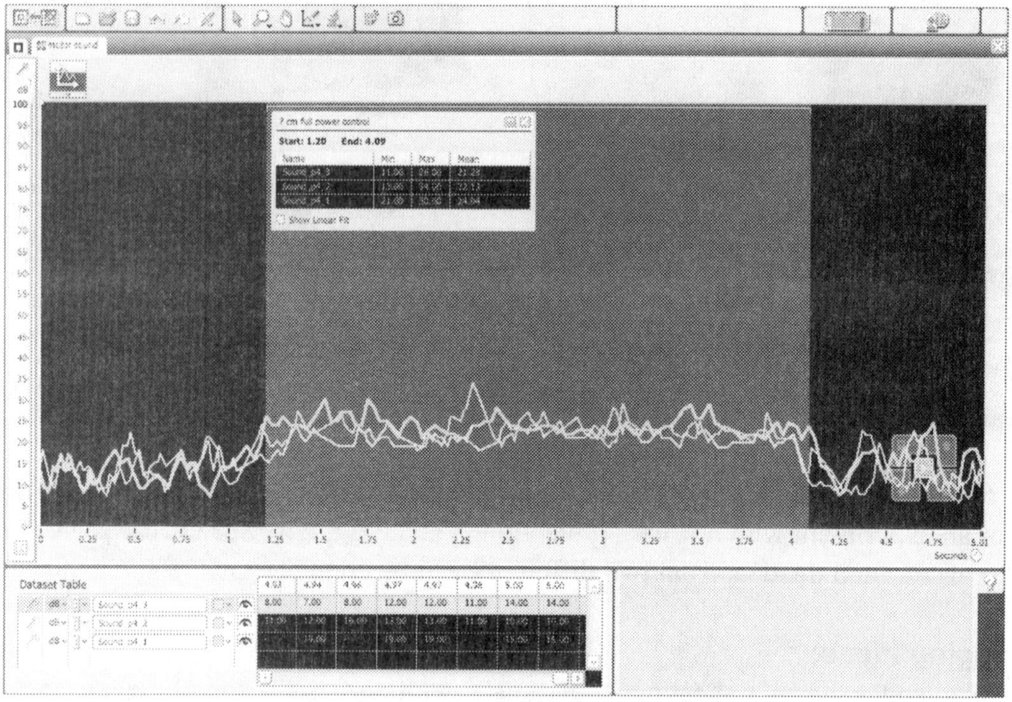

Figure 6.1.5. Sound data from three trials where the sound sensor is placed 10 cm from a motor and the motor is run at full power for three seconds.

Activity Instructions:

1. The goal of this activity is to study the variables that affect the measured loudness of a sound wave. Your first step in such an investigation is to determine which variables you will test. Explore this question with either a simulation (see Pre-Activity Discussion Questions section above) and/or informal experimentation using a set-up similar to Figures 6.1.1 and 6.1.3. What will affect the measured loudness of a sound wave? Refer to the Student Observation and Analysis section of this activity to guide you as you record your results.
2. Pick at least one of the variables that you identified in your experimentation. Design a controlled experiment to document the relationship between this variable and the measured loudness of your source. Be careful with your experiment and do your best to control for other sources of sound in your classroom. Use the dB setting for your sensor.
3. Record the results of your investigation. Refer to the Student Observation and Analysis section of this activity to guide you as you record your results.
4. Share your results with your classmates and document the effect that different variables have on the measured loudness of a sound wave.

5. If time allows create sub groups that tested the similar variables. Determine if the results of these investigations are repeatable.
6. Use the reflection questions of this activity to guide you as you determine what conclusions you can draw from these investigations.

Student Observations and Analysis:

1. Problem definition – Exploring variables that Affect Measured Loudness

Variable	Describe what you did to explore this variable. How did it seem to affect the measured loudness of the sound

2. Chosen variable to investigate: ____________

Variable Values below	*NXT Sound Level (dB)*[3]	*Notes from your experiment*

[3] We recommend using the dB mode for the NXT sound sensor. If, however, you seek to compare the NXT Sound Sensor's measurements with that of the human ear, then use the dBA mode.

3. Now draw a graph of NXT Sound Level versus your chosen variable.

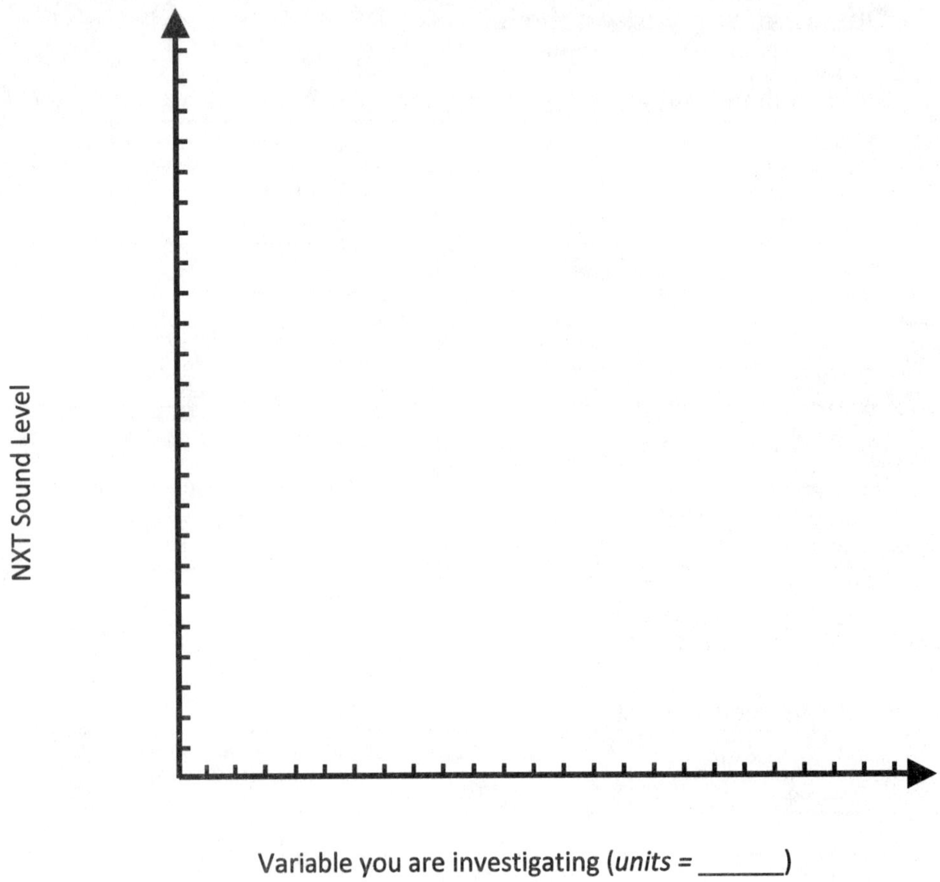

Reflection Questions:

1. What is meant by a sound's "loudness"?
2. How does a sound's "loudness" relate to the amplitude, energy, power, and intensity of the wave?
3. Throughout the activity, we used the phrase "measured loudness". Is there a difference between the loudness of a sound and its measured loudness?
4. What variable did you test in this activity and how does that variable affect the measured loudness of a sound?
5. Did anyone else in your class test the same variable? Were their results consistent with yours?
6. What explains your results? In other words, what explains the relationship that you discovered between your variable and the measured loudness of the sound?

Notes and Extensions:

1. Sound experiments are always a challenge to do in a full classroom of students. Three techniques that work to help manage this challenge are a) coordinate data collection so that the room is quiet during data collection, b) use a separate room for the sound experiment, or c) use a loud source like the piezo buzzers mentioned in this activity.
2. One variable the students may not think to test is the sensor itself. Given identical conditions, do all NXT sound sensors report the same readings?
3. If you have the time, it is a worthwhile exercise for students to try and reproduce each other's results.
4. We have not tried this idea but we thank Bill Kearns for sending it to us and we feel it is a good one to share. Students can explore resonance with the NXT sound sensor. Attach the sensor to a tuning fork. Ring tuning forks of different frequencies at a set distance from the sensor. Ring a tuning fork of the same frequency. What do you observe? Do observations match with a prediction based on resonance?
5. Explore the practical uses of these studies. How can the results of these studies be used in robotics projects?

6.2 Wave Intensity versus Distance

Age Group: 14+ years

Essential Question:

How does the intensity of waves such as light brightness vary with distance from the source?

Recommended Time: 40- 50 minutes

Materials:

Lego Car (RCX/NXT) with a rear light sensor pointing back and a rotation sensor attached. Point source of light (a small, bright bulb able to transmit in all directions is best).

Activity Overview:

This activity investigates the inverse square law for wave propagation. Using the uniform motion of a LEGO car away from the source, a graph of intensity versus distance can be displayed and analyzed. Using the sensors for the NXT, both light and sound can be investigated.

As an introduction, a discussion should be had of the propagation of waves from a point in all directions and the resulting decrease in energy as the wave moves outwards. Models of ripples in water can be used.

Learning objectives:

1. Students use wave models to explain the transfer and transformation of energy.
2. Design and perform controlled investigations.
3. Refine investigations after evaluating variations and inconsistencies.
4. Illustrate and analyze the inverse square mathematical relationship.

Building Ideas:

A LEGO car with fixed axles will ensure linear motion. One motor is sufficient. The objective is to get the car moving at constant speed so that distance and time for light readings are varying equally.

Figure 6.2.1: Fixed Axle car photos

Programming Ideas:

You can use ROBOLAB Inventor and the Front Panel Experiment available on website: www.physicswithrobotics.com or Investigator for this. Mindstorms 2.0 also works well.

The program involves data logging both light intensity and number of rotations (distance) as the car moves slowly away from the light.

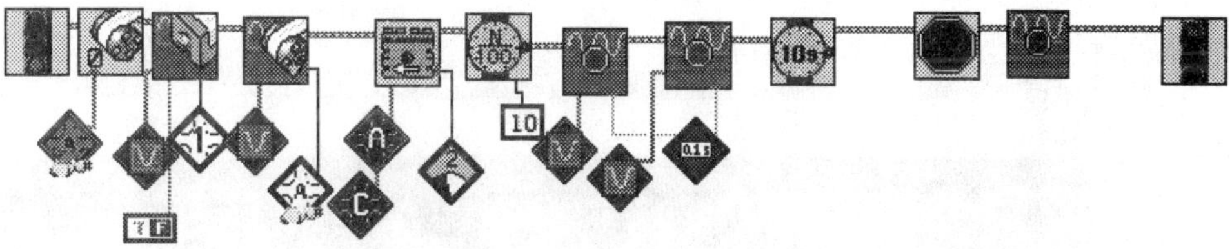

Figure 6.2.2: Robolab program.

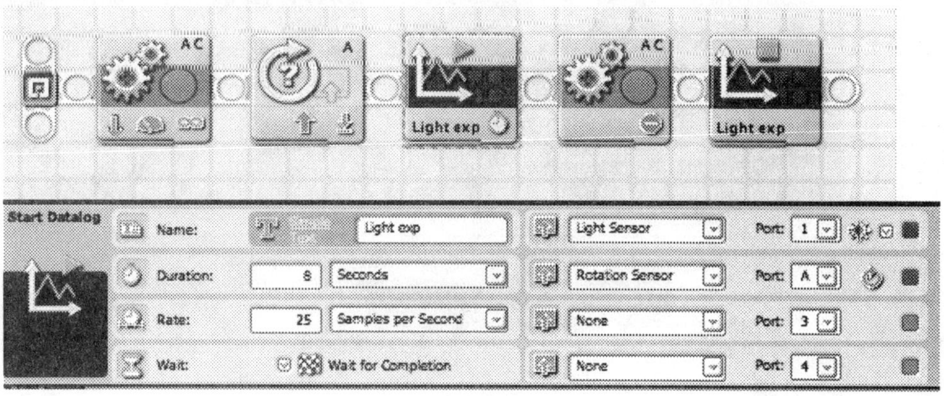

Figure 6.2.3: Mindstorms 2.0 program

Activity Instructions:

1. After downloading the program, set up the light source and car so that the room has, as little external light as possible.
2. Start not too closely to light, so that intensity is less than 100%.
3. Run the car so it moves away slowly in a straight line, still in sight of the light.
4. Data can be uploaded using Mindstorms, Robolab Investigator or using Front Panel Indicators.
5. Graphs of rotation angle and light Intensity versus time can be displayed. If possible combine these to graph Light Intensity versus number of rotations.

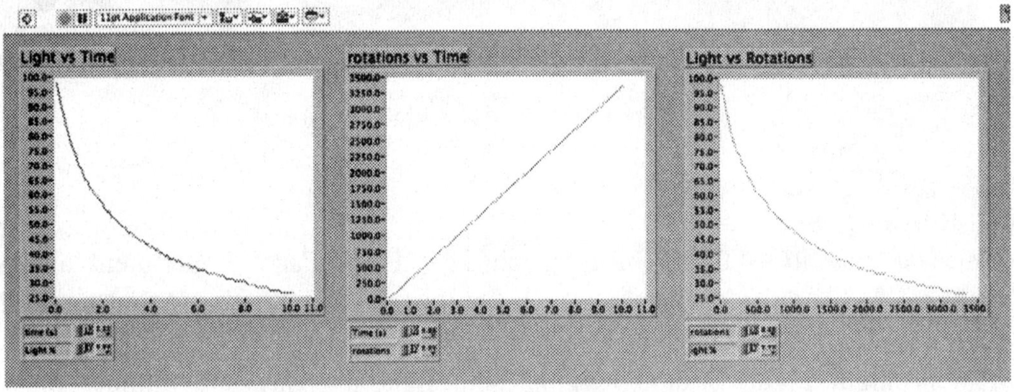

Figure 6.2.4: Front Panel Sample graphs

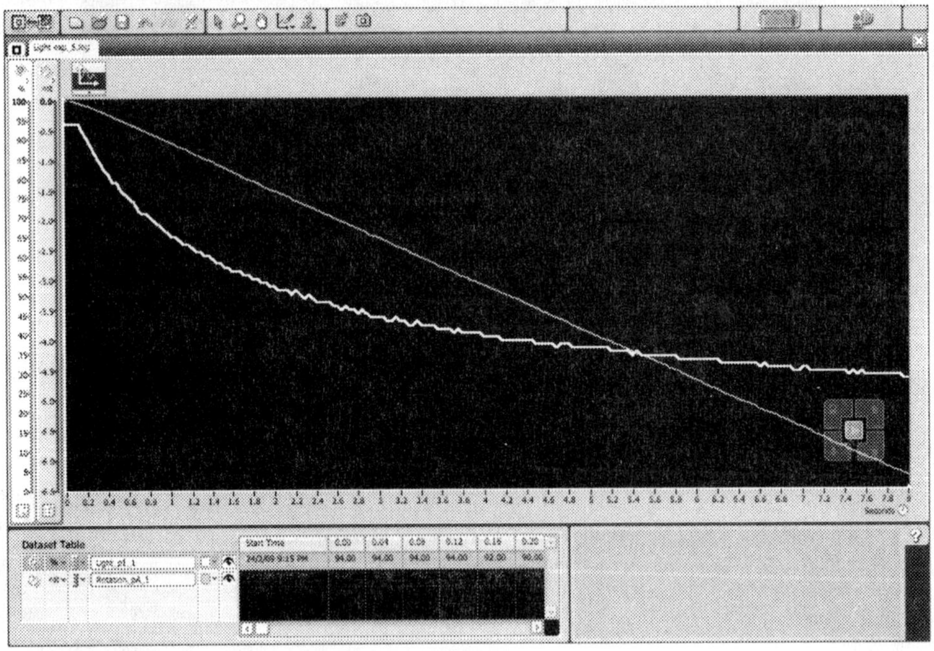

Figure 6.2.5: Mindstorms 2.0 Sample graphs

Student Observations and Analysis:

Observations and Analysis

1. Did the car move uniformly?
2. How did the light intensity vary over time?

Reflection Questions:

1. What can you conclude about light intensity versus distance from the source?
2. If possible determine the relationship between Light Intensity and distance.
3. What factors affected the accuracy of this experiment?
4. Is the NXT light sensor the best for this experiment, if it measures levels of Infrared light?
5. Does changing size and power of the light source have any effect on the results?
6. Will these results be similar for sound waves?

Notes and Extensions:

1. The light graph will not always be smooth due to variations in light emission and reflections around the room, limitations of the sensor and vibrations of the car. The principle of an inverse relationship however, should be seen by the approximation of shape of the curve to a hyperbola where $I \propto 1/d^2$.
2. Find the mathematical relationship between light intensity and distance by saving the data to Excel, Loggerpro or some other graphics program and determine the regression function. Save the Log File and open this file directly from your spreadsheet program. If the file cannot be read, copy and paste the columns into the spreadsheet. The accompanying screenshots and function analysis are taken from the Loggerpro graphing analysis program.

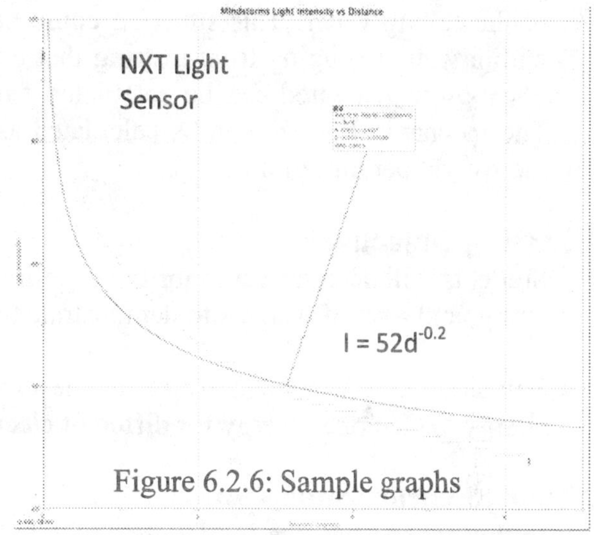

Figure 6.2.6: Sample graphs

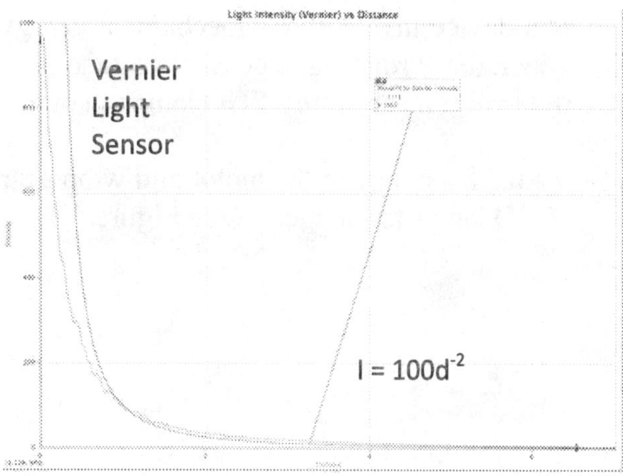

Figure 6.2.7: Sample graphs

6.3 Conversion of Mechanical Energy into Electrical Energy

Activity Level: 14+ years

Recommended Time: 40-50 minutes

Materials:

NXT motor
LEGO lamps and NXT conversion cable
Pulley and a string
Mass with a hanger
Vernier current and voltage probes

Activity Overview:

When falling water rotates, the rotor of the electrical generator electricity is generated. The power generated depends on the rate of change of the potential energy of the falling water. This situation could be modeled by replacing the falling water with a weight hanging from a string that goes over the pulley attached to the motor. The power generated can be calculated from measured voltage and current values. The mechanical power can be calculated as the change of the mechanical energy of the weight per unit of time.

Learning Objectives:

Students will design a setup for the experiments consisting of the NXT motor, pulley, string, and a set of weights to demonstrate the energy conversion.

Students will experimentally study the efficiency of the conversion of mechanical energy to electrical energy for different electrical loads.

Activity Instructions:

In this experiment, the NXT motor will be used as generator. An electrical generator is a device that converts mechanical energy into electrical energy. If the rotor of the generator is rotating, electrical voltage is produced and the current can flow through the load (one or more LEGO lamps in our case).

1. Attach a pulley to the motor and wrap a string around a pulley and connect 1-3 LEGO lamps to the motor (*see* Figure 6.3.1).

Figure 6.3.1: Experimental set-up

Fix the motor to the tabletop or hold it in your hand. Hang a weight from the string and test your system. While the hanging weight moves down, observe that the LEGO lamps emit light. The brightness of the lamp depends on the power dissipated in the lamp.

The power is the rate of producing work $P = \frac{\text{Work}}{\text{Time}}$

The electrical power is defined as a product of voltage (V) and current (I), $P = V \times I$

1. Using Vernier current and voltage probes make a circuit shown in Figure 6.3.2.

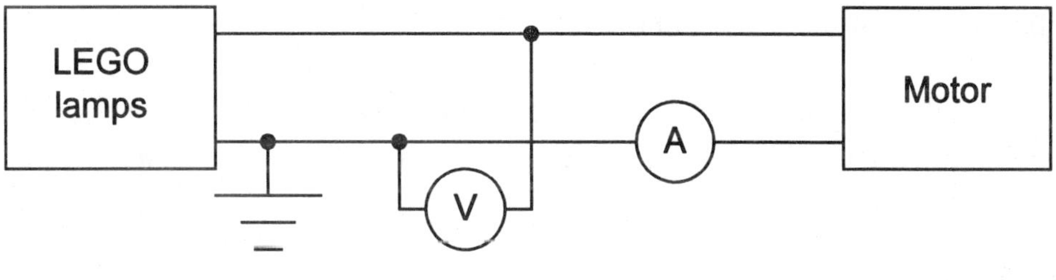

Figure 6.3.2: Electrical circuit diagram of the experiment.

By what means do you produce electrical power delivered to the LEGO lamp?

Rotating the pulley you produce the mechanical work. The faster you produce this work the larger is the mechanical power.

Let's consider the simplest case: when the force exerted on the pulley is constant and the speed of the falling weight is constant. In our set-up, the falling mass will quickly reach equilibrium and therefore fall with a constant speed (terminal speed).

In the case when work is produced due to the constant force applied to the pulley, *work = force × distance (parallel),* and it is called mechanical work.

For the simple case when the object is moving with constant speed through the distance (d), the work (W) produced by the force (F) is $W = F \times l$.

$$\text{Power} = \frac{W}{t} = \frac{mgd}{t}$$

Power can also be calculated as:

$$\text{Power} = \frac{W}{t} = \frac{F \times l}{t} = F \times v \qquad \text{where } v = \frac{d}{t} \text{ is the speed of motion.}$$

In your experiment you will measure mechanical power produced due to the lowering of a weight and you will measure the electrical power delivered to the lamp. The more of the mechanical work converted into electrical work, the larger the efficiency of the process.

2. Connect 1 LEGO lamp and while holding the pulley stationary, connect a weight (1.5 N, for example) to the string on the pulley. Now, release the pulley and let the weight fall. If the weight doesn't move right away, slightly rotate the pulley to start the motion. What can you say about the motion of the weight? Is it moving with a constant speed or with increasing speed? (Hint: Listen also to the motor, the sound is an indicator of how fast it is rotating)

3. Repeat the experiment while collecting quantitative data by measuring current, voltage, distance traveled by the hanging mass, and the time it takes to travel this distance. Record the data in Table 6.3.1. See Figure 6.3.3 with sample data from Vernier current and voltage probes.

4. Connect two LEGO lamps and repeat the experiment. What can you say now about the speed of the moving weight?

5. Connect three LEGO lamps and repeat the experiment. What can you say now about the speed of the moving weight?

6. If the load is connected to the generator, the speed of the motor is increasing from zero to terminal speed of rotation and after that the motor continues to rotate with this constant speed. Therefore, the weight hanging from the pulley will move with constant speed too and the simple equations shown above for power and speed can be used in this case. Calculate the mechanical and electrical power. Calculate the efficiency of the system.

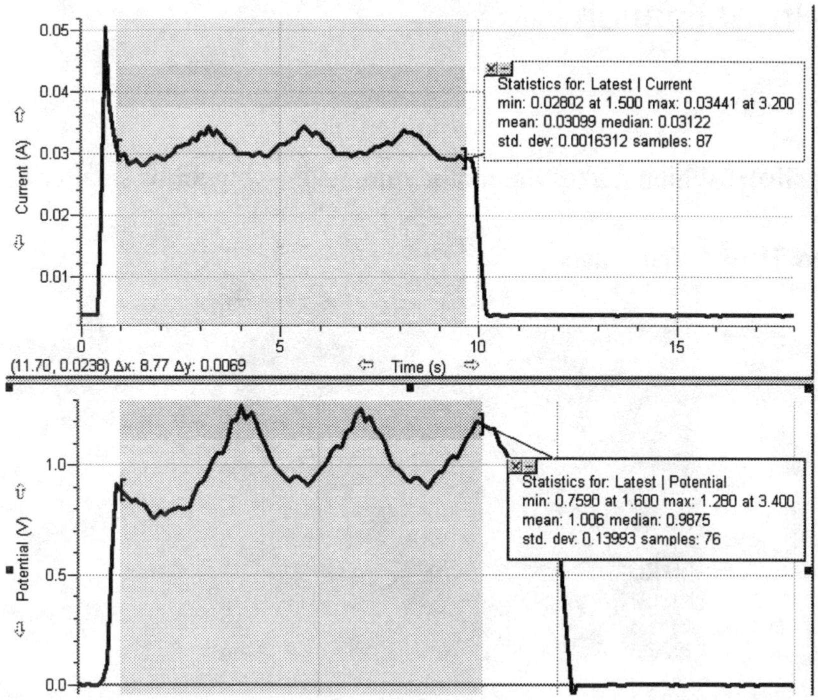

Figure 6.3.3: Current and Voltage produced by the generator.

Using this data, calculate mechanical power developed by the moving mass and electrical power produces by the generator and the efficiency of this conversion of mechanical energy to electrical energy. Record the data in Table 6.3.1.

Student Observations and Analysis:

Table 6.3.1

Weight (N)	Load (# of lamps)	Time, t (s)	Height, h (m)	Work, W (J)	Mechanical Power, P_m (Watt)	Voltage, V (Volts)	Current I (Amps)	Electrical Power, P_e (Watt)	Efficiency (%)
1.5	1								
	2								
	3								

$$Work = work\ of\ force\ of\ gravity = weight \times h$$

$$P_m = \frac{\text{Work}}{\text{Time}}$$

$$P_e = I \times V$$

$$\text{Efficiency}(\%) = \frac{P_e}{P_m} \times 100\%$$

6.4 Battery Investigation

Age Group: 14+

Essential Question: Which makes the motor spin faster – bigger or smaller batteries?

Recommended Time: 30 minutes

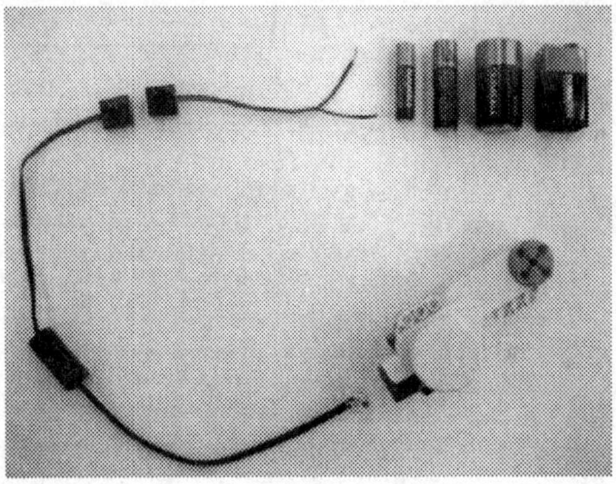

Materials:

LEGO motor (NXT or RCX)
NXT conversion cable if NXT motors are used
Assorted Batteries (e.g. AAA,AA,C,9V)
Multi-meter
Switch (optional)
RCX wire (see building instructions)

Activity Overview:

In this activity students will be connecting a LEGO motor to various batteries to test their assumptions about the effect these batteries have on the speed of a motor. Motors offer students a visual and auditory response to the current supplied by each battery. This lab will challenge their assumptions about batteries and help them build a new understanding about how batteries and circuits work.

Pre-Activity Discussion Questions:

1. What is the difference between voltage and current?
2. What determines how long a battery will last?
3. <u>Prediction:</u> Look at the picture below. Predict the effect the batteries used in this lab will have on the speed of the LEGO NXT motor. *Use the diagram below and label each battery in the drawing: 1 for fastest and 4 for slowest.*

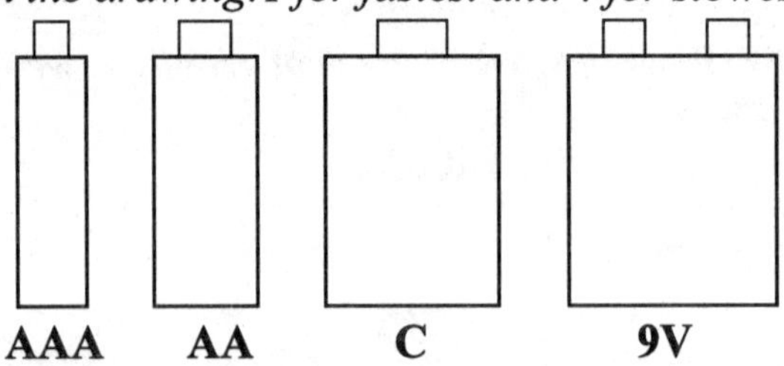

WARNING!
DO NOT USE BATTERIES WITH A TOTAL VOLTAGE OVER 9 VOLTS! In the circuit used in this lab, voltages over 9 volts can lead to currents, which are harmful to you and the motors.

Learning Objectives:

By the end of this activity students will be able to:

1. Define voltage
2. Define current
3. Measure electrical current in a circuit
4. Identify a trend in the relationship between voltage and current
5. Define battery capacity
6. Identify battery design features that affect its capacity and the current produced in an electrical circuit.

Building Ideas:

To connect your LEGO Motor to the batteries, you will need a modified RCX cable. The following steps document how to modify the RCX cable to work in this and other lab investigations involving electricity measurements.

1. Measure your RCX cable and look at the two images below. Cut the cable such that both ends are long enough to use in your circuit. Separate and strip the cable (*see* Figure 6.4.1).

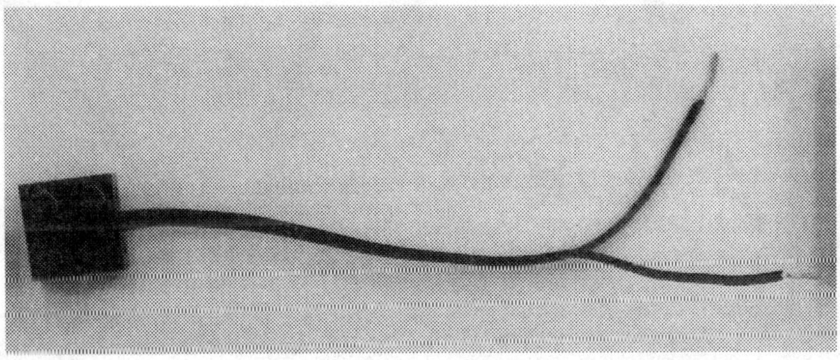

Figure 6.4.1. Modified RCX cable.

2. Repeat step #1 and make three modified cables to build the circuit shown in Figure 6.4.3. NOTE: on the next pages we show an *alternative step* that allows you to build the circuit while only modifying two RCX cables.

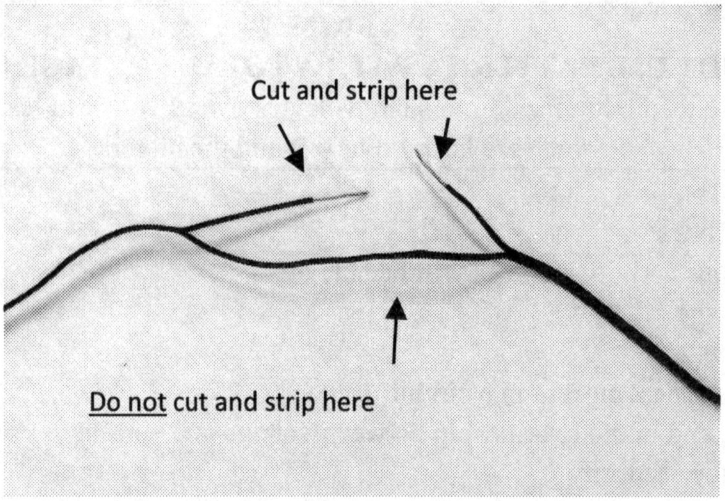

Figure 6.4.2. Another modified RCX cable. This one is well sustained for current measurements.

3. *Alternative Step* – It is possible to make the circuit diagram by modifying only one of the two parallel wires in the RCX cable (Figure 6.4.2. Using a sharp knife, carefully separate the two parallel wires for a distance of approximately 3 cm. Cut and strip the ends of one of these wires. This will allow you to connect your multi-meter and complete the electric current measurement necessary for this lab.

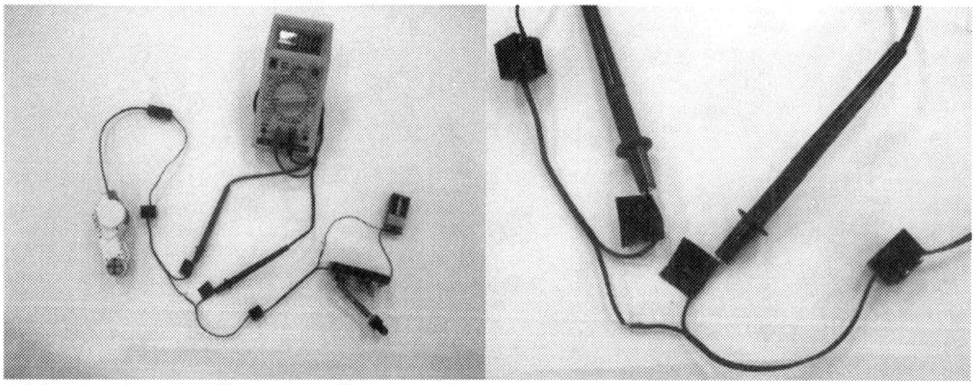

Figure 6.4.3. Experimental design used during this activity. Here, the multi-meter is being used to measure the current in the circuit. Note, this experimental set-up works equally well for the NXT motor.

Activity Instructions:

1. The goal of this activity is to test your predictions about the affect each battery will have on the speed of a LEGO motor. To do this, test each battery separately.
2. Measure the voltage of each battery and record it you observation table. Do this for each battery before moving on to the next step.
3. Connect the battery, motor, multi-meter, and optional switch as shown in Figure 6.4.3.
4. Turn on the motor by closing the switch or connecting the final wire to the battery. Observe the amount of current flowing through the circuit. Write this down in your observation table.
5. Observe speed of the motor compared with the other batteries you test. A simple method for measuring the rotational speed of the motor is to attach a gear or some other LEGO piece that allows you to easily count the # of rotations in a unit of time.

Student Observations and Analysis:

Battery	Voltage Measured	Current Measured	Observation of Motor Speed (revolutions per unit of time)
AAA			
AA			
C			
9V			

Reflection Questions:

1. Based on what you observed, rank order the batteries in terms of which one made the motor go slowest and which one made it move the fastest.
2. Did the size of the battery impact the speed of the motor? Does the size of the battery have any effect at all?
3. What effect does the voltage of the battery have on the current moving in the circuit? What affect does it have on the motor's speed?

Notes and Extensions:

1. If you want to make it easier to count motor rotations, attach a LEGO Beam to the motor.
2. An excellent online simulation related to basic DC circuits can be found at: http://phet.colorado.edu/simulations/sims.php?sim=Circuit_Construction_Kit DC_Only. While this simulation does not include motors, it does support student exploration and thinking about DC circuits in general.
3. For an excellent analysis of several types of LEGO motors, navigate to Philippe Hurbain's website: http://www.philohome.com/motors/motorcomp.htm.
4. By measuring the electric current and voltage of a motor, you can calculate its power. (*Power = Current x Voltage*). Knowing the power consumed by the motor and measuring the mechanical power produced by the motor (*Power = Work / Time*), you can calculate the efficiency of a system (*Efficiency = Power produced / Power consumed*).
5. As an extension to this activity, students can research Ohm's law. They can use LEGO lamps to compare a circuit with motors vs. one with a light. Does Ohm's law accurately predict circuit values for both motors and lights?
6. http://blog.makezine.com/archive/2009/01/electricity_exploration_kit_concept .html

6.5 Investigating the Solar Cell

Age Group: 14+

Essential Question:

How do you get the most out of your Solar Cell?

Recommended Time: 40- 50 minutes

Materials:

Motor, motor connector, LEGO solar cell, NXT/RCX Brick, light sensor, touch sensor and any materials, filters you may want to test.

Activity Overview:

The aim is to compare the light readings of a light sensor to voltage readings of a Solar Cell at various light levels to determine optimum conditions for the solar cell to collect power.

Learning objectives:

1. Model and analyze energy transformation.
2. Design and perform controlled investigations.
3. Refine investigations after evaluating variations and inconsistencies

Pre-Activity Discussion Questions:

1. It is obvious that full daylight in direct sun will generate power from a solar cell. But, what if that's not possible -- how do you find the best spot in a classroom?
2. Where and how can a Solar cell continue to provide adequate power?
3. What factors affect the amount of light getting to a cell and what type of light is required?
4. What effects do different filters have on light intensity?
5. Does reflected light significantly reduce the intensity?
6. How does transmitted light through glass and/or plastic compare?
7. What about artificial light?

Building Ideas:

1. Attach a solar cell to a motor. Check that it works in some light. The motor acts as a resistance sensor. The solar cell when fully functioning uses no power (0V) and must be connected to a load to measure the potential drop and voltage supplied.

Figure 6.5.1: Experimental setup

2. Attach the touch sensor to port 1, motor to port 2, and light sensor to port 3.

Programming:

You can write your own program (see Figure 6.5.2) or use one available from the website at: http://www.physicswithrobotics.com/

Figure 6.5.2: Sample Robolab program: Solar Cell

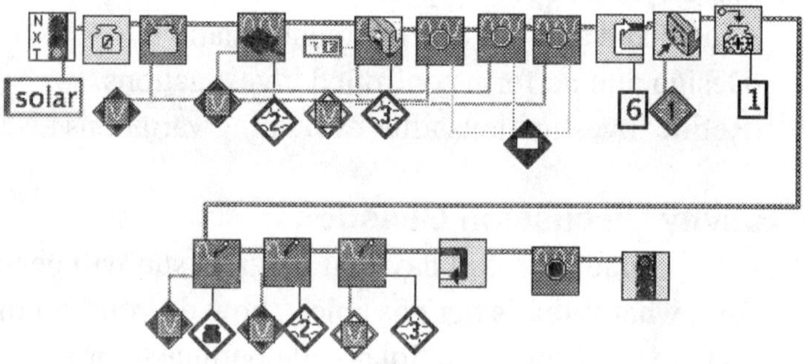

To make this program:

1. Choose Inventor 4 in ROBOLAB 2.9.
2. Data log the voltage across the motor as a generic voltage sensor using the generic icon.
3. Data log the light sensor on port 3.
4. Use free sampling so that the collection is based on the touch sensor on port 1 as the mode for initiating readings.
5. Also data log the number of readings using a container.
6. Take as many data logging readings as required.
7. You can view the data a number of ways:
   a. Using Detective – Data Log View
   b. Using Investigator – copy and paste this program in Program Level 4
   c. Using Front panel – program available from website.

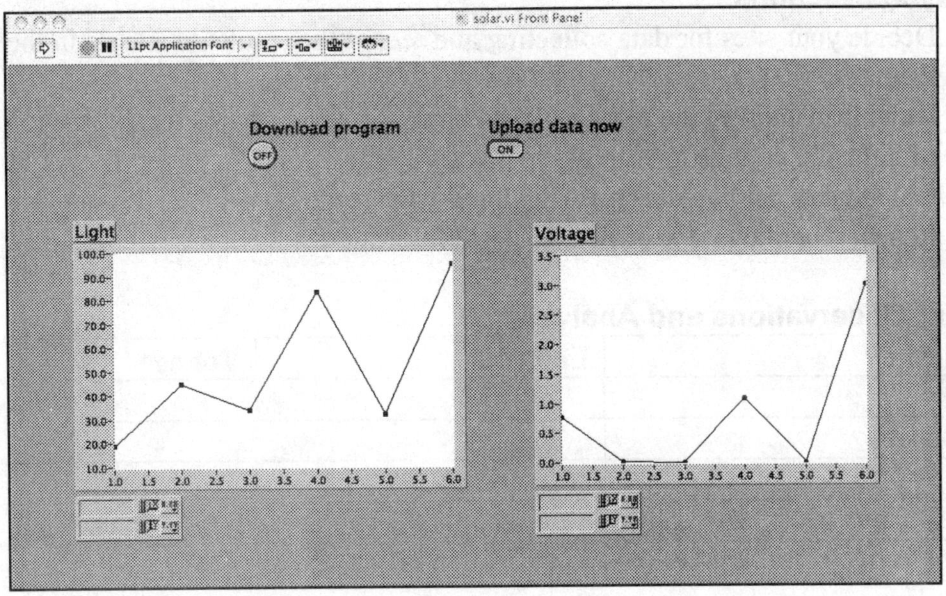

Figure 6.5.3: Sample Front Panel Program Screen for program available from website. Shown here are six different sites where light level and voltage data was collected. Note how different sites yield unexpected results such as high light levels and low voltage!

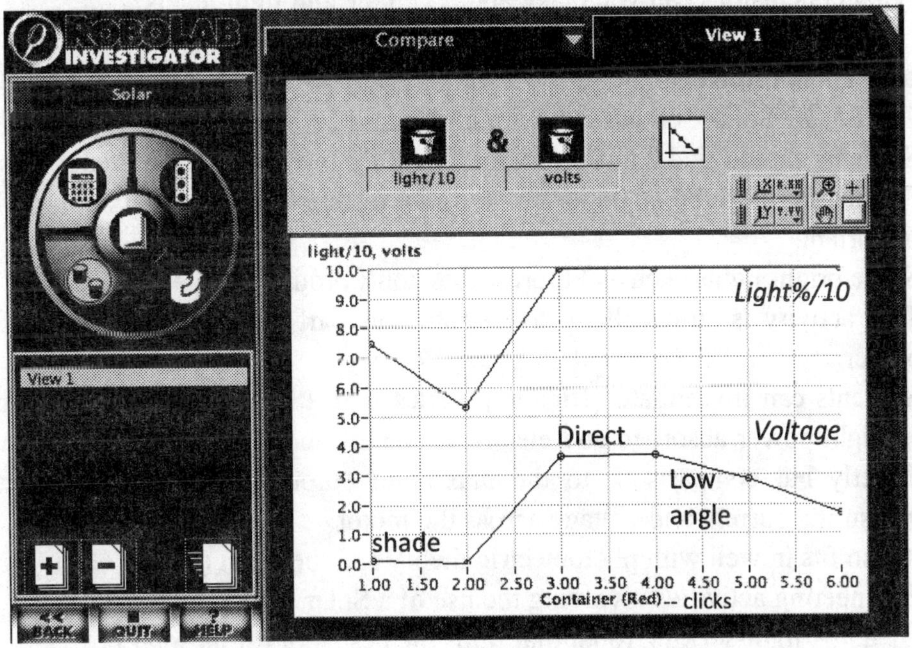

Figure 6.5.4: Data collection and analysis using Investigator in Robolab. The top line represents light levels and the bottom graph represents voltage across the motor. Note the difference here between values take in the shade, with direct light and low angle light.

Activity Instructions:

1. Decide your sites for data collecting and record the results in a table in the order you take them.
2. Download and run the NXT/RCX clicking the Touch sensor each time to record readings of light and voltage in each of the places.
3. Upload data for two data sets (light and voltage).
4. Repeat recordings if necessary to confirm results.

Student Observations and Analysis:

Site	Light %	Voltage

Reflection Questions:

1. Which areas show greatest light values?
2. Which areas show greatest voltage values?
3. Is voltage always in proportion with light level?
4. What other factors are involved?
5. What conclusions can you make about voltage and light levels?

Notes and Extensions:

1. The LEGO Solar Cell has a maximum 3V and 200mA.
2. Students should find a threshold light level for the Solar Cell to generate current.
3. Frequency of light is important too, so testing various filters for optimum colors is important.
4. Some bright lights such as Fluorescents don't produce much voltage.
5. This activity is an excellent investigation as part of an alternative energy unit on solar power.
6. Students can investigate efficiency issues with the solar cell. It is difficult to compare sunlight power absorbed and electrical power produced because of measuring irradiance directly but comparisons to the maximum output power of 0.6W could be done by measuring current and voltage across the motor.
7. It also fits in well with photoelectric theory and the atom (Senior Chemistry or Physics).
8. Engineering activities involving the use of a Solar Cell:
    a. A light-seeking robot that finds the best spot for its solar power.
    b. A motor controlled platform for the solar cell that follows the sun.
    c. A capacitor LED night-light, charged by the solar cell.

6.6 Parallel and Series Circuits

Age Group: 14+

Essential Question:

When a light bulb goes out in your house, do all the other bulbs go out as well?

Recommended Time: 45 minutes

Materials:

3 LEGO motors
3 AA Batteries
1 nine-volt battery
3 NXT conversion cable
Modified LEGO connecting wire
Battery holders

Activity Overview:

In this activity, students will build parallel and series circuits out of LEGO motors and batteries. The motors represent the light bulbs in the question above. Motors offer students a visual and auditory response to the current running in each circuit arrangement. By connecting motors and batteries in parallel and series circuits, students will not only learn the effects a circuit arrangement has on a motor but also the practical application of parallel and series to the electric circuits we every-day life.

Pre-Activity Discussion Questions:

1. What is the difference between parallel and series circuit connections?
2. Using a simulation, build several of the series and parallel circuits listed in activity instruction #1. An excellent simulation can be found at the following website: **http://phet.colorado.edu/simulations/index.php?cat=Electricity_Magnets_and_Circuits.** Click on the "Circuit Construction Kit (DC Only)" simulation.
3. Look at the material list for this activity. Predict which arrangement of batteries and motors will make any single motor move the fastest and which arrangement will make any single motor move the slowest.

Learning Objectives:

By the end of this activity students will be able to:

- Build parallel and series circuits
- Explain the effect circuit design (i.e. Parallel or series) has on the performance of both motors and batteries

Building Ideas:

To connect your LEGO Motor to the batteries, you will need a modified RCX cable. The Chapter 6 -- Battery Investigation activity in this book documents how to create this cable. The following figures document how to use this cable to create series and parallel connections between a nine-volt battery and the NXT Motors. The same connections can be made with RCX motors as well.

As an example, we are showing three motors connected to a nine-volt battery. Figure 6.6.1 shows both the series and parallel wiring arrangements.

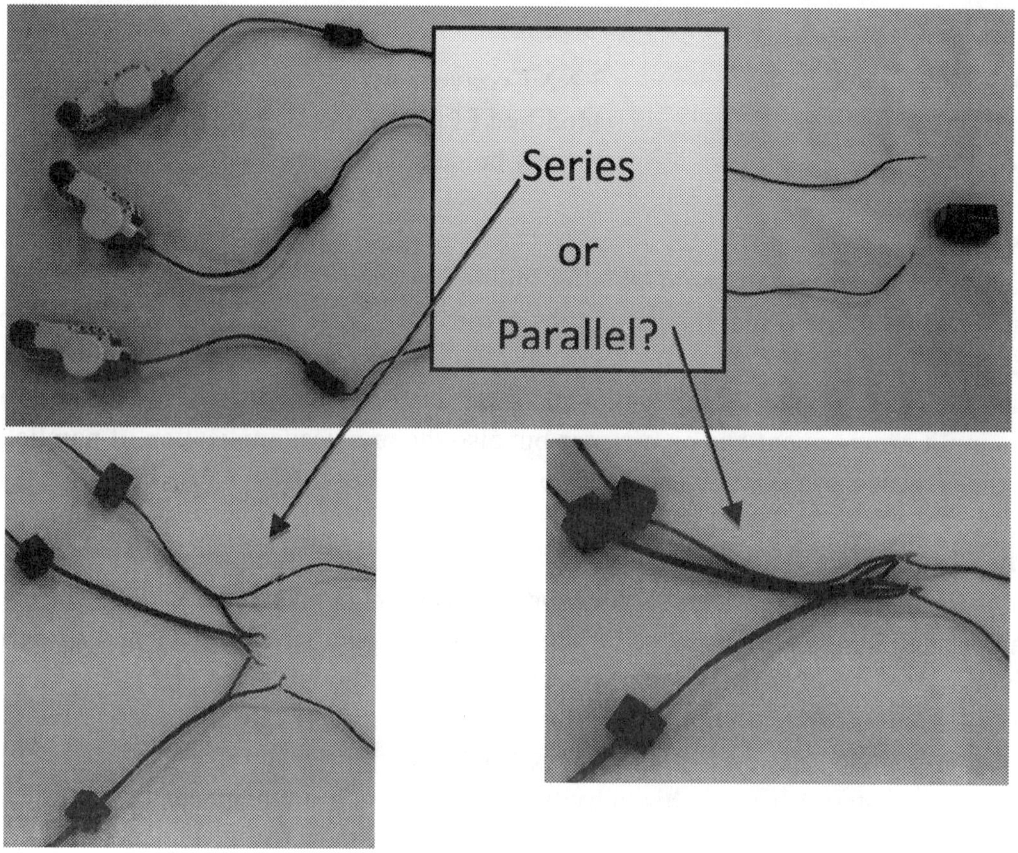

Figure 6.6.1. Series and Parallel connections between a battery and LEGO motors. This wiring works equally well for NXT and RCX motors.

Activity Instructions:

1. On a separate sheet of paper, draw circuit diagrams (schematics) for the following circuits:
    a. 1 motor and 1battery
    b. 2 motors in series with 1battery
    c. 3 motors in series with 1battery
    d. 2 batteries in series with 1 motor
    e. 3 batteries in series with 1 motor
    f. 2 motors in parallel with 1 battery
    g. 3 motors in parallel with 1 battery
    h. 2 batteries in parallel with 1 motor
    i. 3 batteries in parallel with 1 motor
2. Look at Table 1 in the student observation section of this activity. These are the circuits you must build and observe. Build each circuit in Table 1 using the appropriate battery, circuit arrangement, and LEGO motor(s). Measure and record motor speed[4] as a way to document the effect that the circuit arrangement has on the motor's performance.

WARNING: **DO NOT USE BATTERIES WITH A TOTAL VOLTAGE OVER 9 VOLTS!** In the circuits used in this lab, voltages over 9 volts can lead to currents, which are harmful to you and the motors.

[4] A simple method for measuring the rotational speed of the motor is to attach a gear or some other LEGO piece that allows you to easily count the # of rotations in a unit of time.

Student Observations and Analysis:

Observation Table One

Circuit Description	**Observation of Motor Speed (revolutions per unit of time)**
1 motor and 1 nine-volt battery	
2 motors in **series** with 1 nine-volt battery	
3 motors in **series** with 1 nine-volt battery	
1 AA battery and 1 motor	
2 batteries in **series** with 1 motor	
3 batteries in **series** with 1 motor	
2 motors in **parallel** with 1 nine-volt battery	
3 motors in **parallel** with 1 nine-volt battery	
2 AA batteries in **parallel** with 1 motor	
3 AA batteries in **parallel** with 1 motor	

Reflection Questions:

1. Summarize your findings about connecting additional motors in series with a single battery.
2. Summarize your findings about connecting additional batteries in series with a single motor.
3. Summarize your findings about connecting additional motors in parallel with a single battery.
4. Summarize your findings about connecting additional batteries in parallel with a single motor.
5. What are the advantages and disadvantages of connecting motors in series?
6. What are the advantages and disadvantages of connecting batteries in series?
7. What are the advantages and disadvantages of connecting motors in parallel?
8. What are the advantages and disadvantages of connecting batteries in parallel?
9. How do the observations you made in this lab relate to circuits that you interact with in everyday life? Explain at least two examples.
10. Why do you think that two different types of batteries were used in this activity? Why not simply use just one type of battery?

Notes and Extensions:

1. We used the motor rather than the lights because a motor's response to current is both clearly seen and heard. A light bulb's response can be hard to distinguish when without measuring current and voltage levels. With a motor, students can learn a great deal about the difference between parallel and series circuits without making current and voltage measurements. Note, however, that the LEGO lamps can certainly be used in this activity. Make sure you do so in a way that makes their response clearly visible. Creating black paper hoods works very well for this purpose.
2. Measure motor current and voltage for each motor in the circuits built for this activity. Are the results consistent with Ohm's and Kirchoff's laws?
3. Devise a battery draining experiment with rechargeable batteries. Predict which will drain a battery faster – motors in series or motors in parallel.
4. Using Investigator or Mindstorms 2.0 and Spreadsheet and Vernier sensors datalog Voltage versus current as batteries are changed.
5. http://blog.makezine.com/archive/2009/01/electricity_exploration_kit_concept.html

Chapter 7

Challenges

7.1 NXT Distance Match Problem Solving Challenge

Age Group: 15+ years

Challenge:

Make a LEGO car match a motion graph constructed by your partner involving up to 3 steps with varying speeds (positive, stop and negative for varying times).

Recommended Time: 40 – 50 mins

Materials:

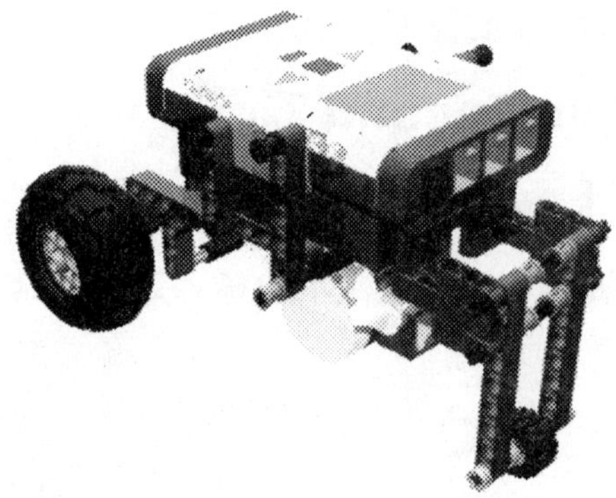

Use an NXT LEGO car with at least one motor encoder.
MINDSTORMS[1]/ROBOLAB Software.

Activity Overview:

Work in groups of two. The Challenge is in two parts:

1. Generate a three-stage motion graph (Distance versus Time) with a car for your partner to analyse.
2. Analyse your partners graph and reproduce it with your car.

You will need the speed to power level conversion! This challenge uses your findings from the Activity: "4.2 An Investigation of NXT Motor Speed"

Learning Objectives:

1. Apply physics concepts to construct and analyse stages of motion.
2. Determine speed from a distance – time graph.
3. Applying a known mathematical relationship to a problem.

LEGO Sample Building Instructions:

You can use any NXT car (one motor is sufficient).

Programming:

The following sample was done using the Front panel of ROBOLAB. See Figure 7.1.1. A download of the program "Distance match front panel.vi" is available at: http://www.physicswithrobotics.com/

---

[1] This activity can be done with the MINDSTORMS 2.0 software. See our website for instructions – PhysicsWithRobotics.com.

Sample Front Panel Window:

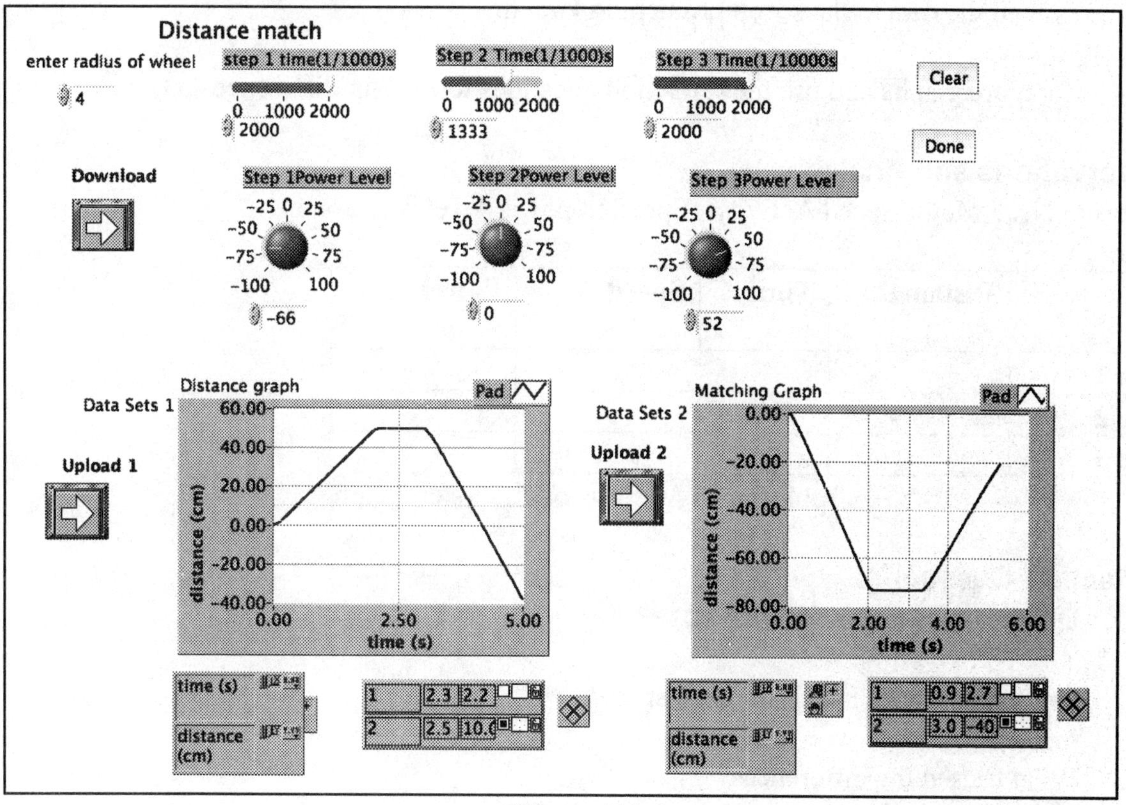

Figure 7.1.1

Activity Instructions:

Part 1

1. Using the Front Panel Controls, generate a Three Step Distance Graph for your partner to analyse and match.
2. This is done by entering time and power level, downloading to your car, running it and then uploading the data.
3. Reset the controls. You don't want to give your partner any help!

Part 2

1. Now analyse your partners graph by determining power levels and times required from the graph.
    a. **Note:** You will need to convert the distances shown in centimeters to a power level. If you have completed the Chapter 4 Data logging activity "4.2 An Investigation of NXT Motor speed" you will have the direct conversion of speed to Power level.

2. When power levels and times are determined enter the data and run your car.

3. Upload the data to the second Matching Graph.

4. Compare graphs and use the reflection questions to explain differences.

Observations and Analysis:

From NXT Motor Speed Activity: Speed to power level conversion = ______

	Distance	Time	Speed	Power level required
Step 1				
Step 2				
Step 3				

Table 7.1.1 Analysis of three-step graph

Reflection Questions:

1. Did the times match for each stage?

2. Did the distances match for each stage?

3. What caused the differences?

Notes and Extension:

1. The students need to have determined speed from distance - time graphs.
2. The activity of "4.2 An Investigation of NXT Motor speed" is necessary before this challenge.
3. It is important students can work together to challenge each other.
4. If computer numbers are limited they can work together on one computer and swap the data with another group.
5. Follow up extension activities to this include:
   A. Matching Velocity - Time graphs
   B. Matching Acceleration Time graphs
   C. Do a speed control activity such as a "Traffic Zone". For example consider a "school zone" where the model car would travel at 1 m/s then slow down to 0.5 m/s for 1 meter past a model school. The same could be done for a "construction zone or any other type of "traffic zone".

7.2 Smooth Ride

Age Group: 14+

Essential Questions: What is the best way to transport a full canister of water as fast as possible over a bumpy road?

Recommended Time: 90+ minutes

Materials:

NXT or RCX Robotics kit
Small canister and overflow area
System to measure acceleration (accelerometer or motion detector data that can be analyzed for acceleration)

Activity Overview:

In this activity, students will design a LEGO system to transport a small cup of water over a bumpy path. To help students obtain feedback on their design before the final trials with water, acceleration measurements will be taken and analyzed. The final goal of the activity is to quickly traverse the course of bumps while retaining the most water in the cup.

Pre-Activity Discussion Questions:

1. What is the difference between speed and acceleration?
2. How do words such as "smooth" and "bumpy" relate to the speed, acceleration, and forces of a car driving on a rough-road.
3. When driving a car on a rough road, how does speed influence the smoothness of the ride?
4. If possible, watch and discuss the Discovery Channel's Mythbusters episode #58 "Rough Road Driving" for an investigation into the relationship between speed and acceleration when driving over washboard roads.
5. What characteristics of cars (and even bikes) relate to the "smoothness" of a ride?
6. How do the definitions and mathematical relationships related to speed, acceleration, inertia, and force help designers of cars and bikes create "smooth rides"?

Learning Objectives:

By the end of this activity, students will be able to:

1. Explore the concepts of speed and acceleration through an applied design challenge
2. Relate their design decisions to the concepts acceleration, inertia, and force.
3. Discuss how Newton's 2nd law of motion connects to our everyday life

Building Ideas:

Control data is important to have in this activity. As with any scientific investigation, control data allows you to determine the impact, changing variables have on the system. If you decide to have the students collect control data using a car that is not specifically designed for the Smooth Ride task, we suggest using any of the designs in Chapter 2 of this book. Alternatively, the "Constructopedia" resource (see resources section in Chapter 2) will also provide excellent control models.

Students will design systems to suit their own ideas. We do not provide examples here. We do provide two tips in this section:

1. Figure 7.2.1 shows an example water canister and overflow container for the water tests used at the end of this activity. Our canister is an item that used to be easy to obtain - a film canister.

Figure 7.2.1 The film canister holds water for the final water tests in the activity. The plastic storage bin is used as an overflow container. We glued bricks to the bottom of the canister so it can be attached to the students' vehicles.

2. Figure 7.2.2 depicts a set of LEGO weights. Used in this activity, they allow students to test the affect of increased mass (inertia) on the acceleration of the vehicle.

Figure 7.2.2 LEGO Weight Elements (part #9936) used to add mass to a vehicle.

Data logging Ideas:

Collecting a vehicle's acceleration data could be an activity in itself. We chose, however, to demonstrate only two acceleration measurements systems in this section of the write-up – a Vernier motion detector example and a HiTechnic Accelerometer example. These two acceleration measurement devices allow us to demonstrate data logging with Vernier's Logger Pro software and Robolab. At the time of writing this activity, we did not have a straightforward method to present for data logging acceleration via the Mindstorms 2.0 software[2]. What we present, however, are not the only two methods for collect acceleration data. Refer to the Notes and Extensions part of this activity write-up for more information regarding acceleration data collection.

Figure 7.2.3

Figure 7.2.3 depicts an acceleration data logging technique using a Vernier motion detector. To collect useful data, we set the data collection rate at 50 samples per second and we raised the sensor up on top of a box.

---

[2] This situation is likely to change with the release of a Mindstorms 2.0 data-logging block for the HiTechnic acceleration sensor. Look to the HiTechnic website for updates.

Figure 7.2.4. illustrates sample graphs from the Vernier Logger Pro Software. Here, we see both "position" and "acceleration" data for vehicles moving at full power and at 25% power. Clearly, the speed of a vehicle driving over a series of bumps has an impact on the acceleration of the vehicle.

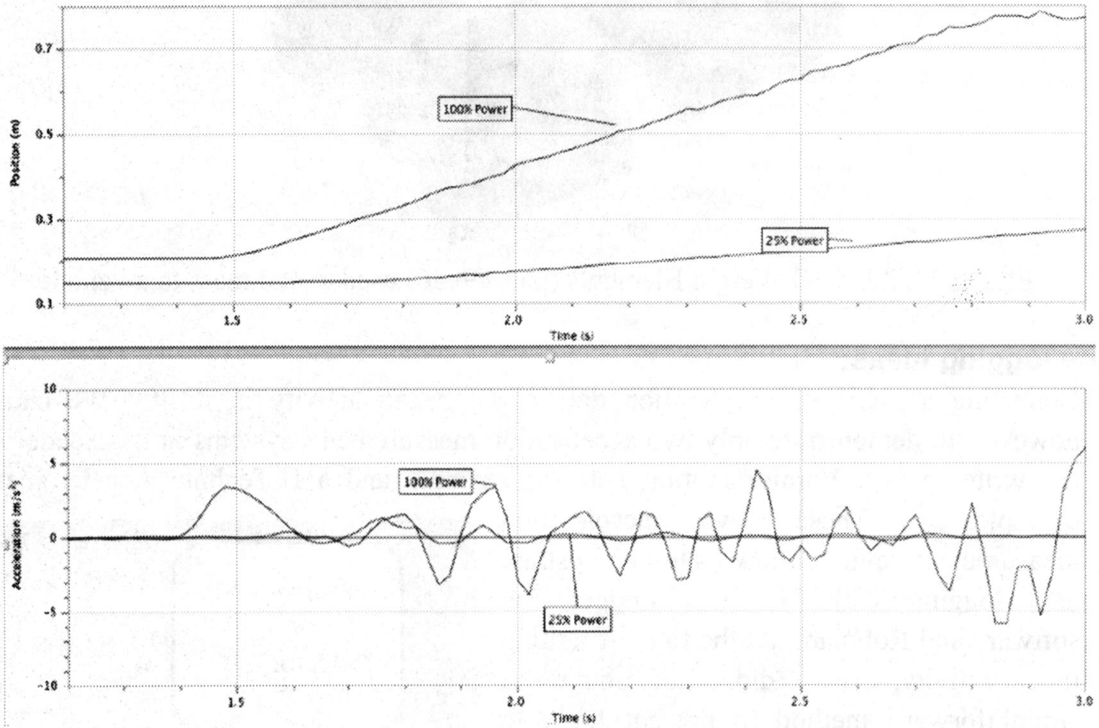

Figure 7.2.4

The second acceleration datalogging method we will share with you utilizes the HiTechnic acceleration sensor and Robolab's Investigator. Figure 7.2.5 shows the sensor in horizontal and vertical positions.

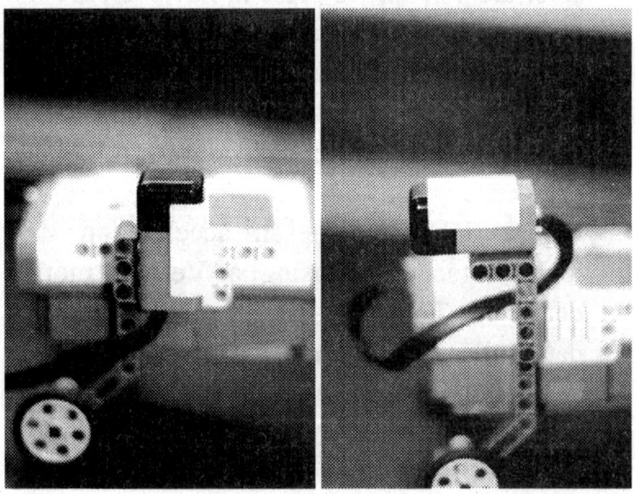

Figure 7.2.5 NXT car and HiTechnic acceleration sensor.

Figure 7.2.6 shows Robolab programming code to capture data from the sensor for three seconds and at a data collection rate of 100 samples per second. Figure 7.2.7 shows example acceleration data of a vehicle driving at 100% power and 25% power.

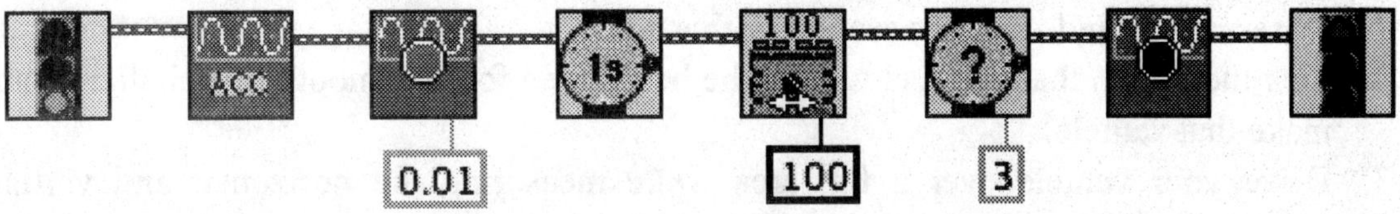

Figure 7.2.6 Datalogging code for the HiTechnic acceleration sensor using Robolab Investigator (ROBOLAB Program Level 4)

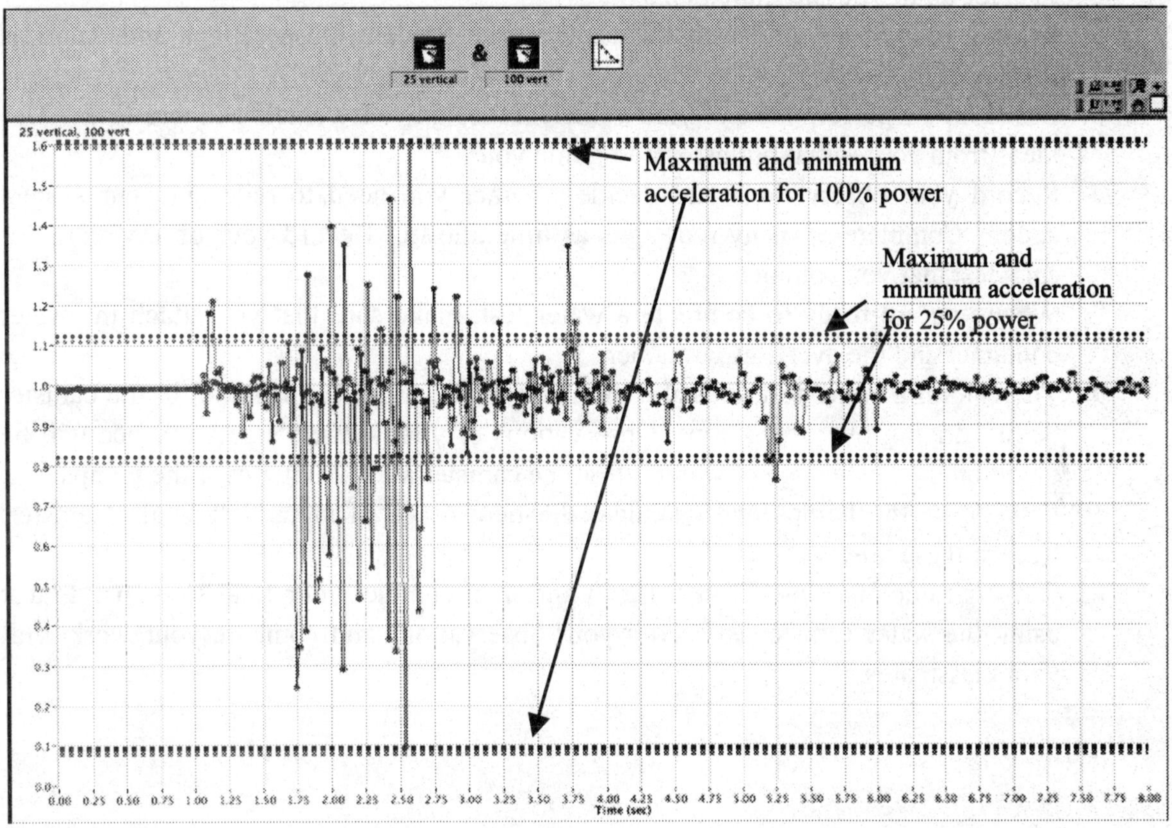

Figure 7.2.7 Example data from the HiTechnic acceleration sensor displayed in Robolab Investigator. To obtain the data, the vehicle moved at power level of 100% for one trial and at a power level of 25% for the other trial.

Activity Instructions:

1. Brainstorm and document at least three different designs of a vehicle that will produce the "smoothest" ride while driving over a series of bumps as fast as possible. At the end of this activity, you will conduct a water test wherein you will carry a full canister of water across the bumps while moving as quickly as possible and keeping as much water in the canister as possible. Document your ideas in the Student Observations and Analysis section of this activity.
2. Pick the design that you feel will be the best suited for the smooth ride challenge and make that vehicle.
3. Drive your vehicle over a flat area *while* measuring the horizontal and vertical acceleration of your vehicle. Do this multiple times to generate a control against which you can compare your subsequent runs.
4. Drive your vehicle over the bumps course *while* measuring the acceleration of the vehicle. Measure both the vertical and the horizontal accelerations. Do multiple runs for all of your measurements.
5. Record your observations on the chart in the Student Observations and Analysis section of this activity.
6. Based on the performance of your vehicle, redesign your vehicle and test again. Use ideas from your initial brainstorm to guide you.
7. Record your observations and decide whether you need to redesign your system again. Complete as many redesigns as time allows. Record your observations for each test that you complete.
8. When you are ready to complete a water test, make sure that you attach the water container and the overflow container.
9. To perform a water test, measure the mass of the water (or volume) in the canister before driving over the bumps. Observations of your vehicle's performance will be based on the percentage of water left in the canister after driving over the bumps.
10. Drive over the bumps and calculate the percentage of water left in the canister. Record these observations.
11. Redesign one last time (if necessary) and then conduct your final smooth ride test using the water canister. Record your observations and compare your work with your classmates.

Student Observations and Analysis:

Design Ideas and Testing

Use sketches and descriptions to document your ideas.	Explain why you believe your ideas will work.	Summarize the results of evaluation measurements. (i.e. how fast did your vehicle travel and what did the acceleration data look like?)	Describe any changes you will make to your next design
Design #1			
Design #2			
Design #3 *If you need additional space to document the results of tests on additional designs, use extra paper and attach it to your lab write-up.*			
Design used for the water test		*Describe the results of your water test. What percentage of the water remained in the canister? How fast did your vehicle move?*	
Final Design			

Reflection Questions:

1. Which of your robot designs yielded the lowest accelerations as measured by an accelerometer or motion detector?
2. Did low accelerations in your trials predict a smooth ride when you tested your robot with the water?
3. Observe the performance of your classmates' designs. Describe the best designs for achieving low accelerations. Explain why these are the best designs.
4. Describe the best designs for achieving a smooth ride as measured by the water test. Are these the same designs you documented in the previous question? Explain why these designs worked well for the water tests.
5. How did the "speed" aspect of this activity affect your design decisions?
6. Did you test weight of the car? If so did this affect the smoothness of the cars ride?

Notes and Extensions:

1. You could spend a great deal of time exploring the tools to measure acceleration in this activity. A few thoughts to consider as you prepare to do this lab:
    a. Pick the acceleration measurement system based on the availability of equipment. HiTechnic, Mindsensors, Vernier, Pasco, and Fourier all have excellent tools for measuring acceleration.
    b. You will pick from two measurement strategies: using a "motion detector" (i.e. an ultrasonic range finder) or an accelerometer.
    c. The NXT's distance sensor is an ultrasonic range and *could* be used in this activity. At the time of this writing, however, the process of completing the calculus analysis on the position data to obtain acceleration data was beyond the scope of the book. If you would like to try this, we recommend either writing a custom Robolab program to analyze the position data (use "Compute Tools 5") or exporting the data and importing it into a software package capable of completing the calculus analysis.
    d. Do not overlook other acceleration measurements systems. Accelerometer chips are very cheap. If you have the inclination and time, you can learn to interface an accelerometer to a BASIC Stamp, PIC, or Arduino microprocessor and then create your own custom output to represent the data.
    e. Investigate cell phone applications. For example, the Apple iPhone has a few programs that can display graphs of the device's accelerometer data. Your students may be walking into the classroom with great acceleration measurement tools!
2. Engineering design challenges such as the one presented in this activity offer the opportunity for performance based assessment. One grading system we have used for activities similar to Smooth Ride is to split the grade between *Process* and *Performance*. For example, two thirds of the grade could be derived from the documentation of the process they used to design a successful system – i.e. their

work on design, observation, redesign and reflection questions. The remaining one third of the grade would be derived from their performance on the final water test. Points could be awarded for speed above a certain level and for the percentage of water left in the canister. By including performance assessments, you give students a clear goal to work toward while making a valuable point about the importance of measuring performance in real world systems. Here the percentage of water left in the canister allows for a concrete and visual performance indicator.

7.3 Ten Second Timer

Age Group: 16+

Essential Question: Given some LEGO pieces and a LEGO weight, can you make a simple device that accurately times 10 seconds?

Recommended Time: 90 – 180 minutes

Materials:

LEGO beams, plates, pegs
NXT or RCX
LEGO weight
Light sensor

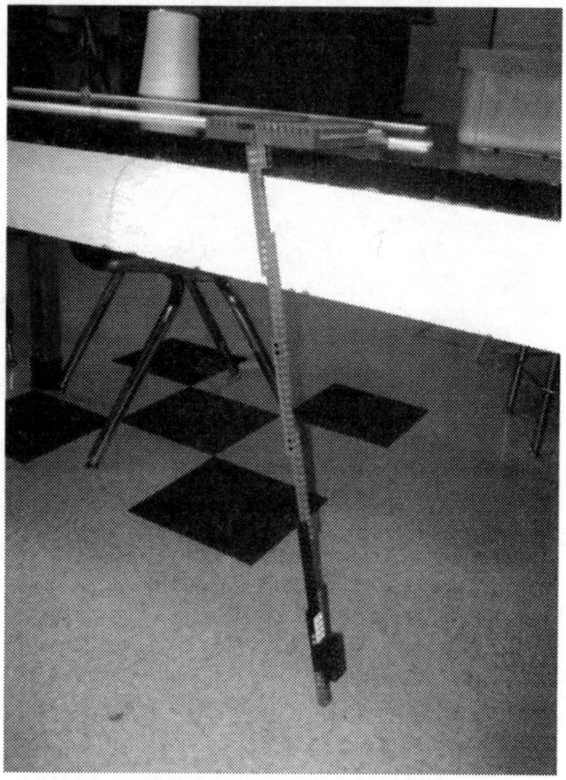

Figure 7.3.1 sample 'LEGO Pendulum'

Activity Overview

In this activity, students will determine how well the equation for a simple pendulum predicts the period of a physical pendulum constructed from LEGO Beams and LEGO weights. They will do this by:

1. Designing a LEGO Beam physical pendulum that accurately times ten seconds with a whole number of swings.
2. Building their "ten second timer."
3. Finally, using the sensors from an RCX or NXT kit, students will create a system that tests the accuracy of a "10 second timer".

Pre-Activity Discussion Questions:

1. How are pendulums used in clocks?
2. As a pendulum's swing (amplitude) becomes smaller, what happens to its period?
3. What is the difference between a *physical pendulum* and a *simple pendulum*?
4. What is the mathematical relationship that determines the period of a simple pendulum?
5. What does "center of mass" mean and how can you determine the center of mass for a LEGO Beam physical pendulum?
6. How can you use the relationship for determining the period of a simple pendulum and a working knowledge of "center of mass" as a starting point for making a prediction regarding the period of a LEGO Beam physical pendulum?

Learning Objectives

By the end of this activity students will be able to:

1. Use an idealized system (simple pendulum) to make predictions about a real world system (physical pendulum).
2. Identify the location of a center of mass of a system.
3. Refine predictions about a systems behaviour based on careful observations.
4. Predict characteristics of a physical pendulum and engineer a 10 second timer.
5. Create a system that tests the accuracy of a 10 second timer.
6. Identify sources of human error and create a system that minimizes or eliminates human error.

Building Ideas:

In the three parts of this activity, students will build two systems. They will build physical pendulums and an RCX or NXT based system to measure the performance of a physical pendulum. We share here, ideas related to both systems.

Figure 7.3.2 shows the pieces of a basic physical pendulum system made from LEGO parts. In the pictures, you will notice four key parts to the system:

1. LEGO beams
2. LEGO Weight Element (part #9936)
3. Top plate and support to keep the pendulum stationary
4. Axle for pendulum rotation

Figure 7.3.2 Physical pendulum created from LEGO parts.

Figure 7.3.3 shows an example NXT system for measuring the performance of the activities culminating challenge – the "ten second timer". Here, you see the pendulum being tested, the NXT with a light sensor, and a piece of paper behind the pendulum. The paper serves as a highly reflective surface to improve the light sensor readings. The way this testing system works is as a "break beam" type system. The NXT uses the light sensor to count the number of times the pendulum has swung.

Figure 7.3.3 A "break beam" type counting system.

After a pre-determined number of swings, the NXT displays the time. For instance, if students built a pendulum to time ten seconds with 5 swings then this system would count five swings and display the time. The time displayed gives students clear feedback on the accuracy of their pendulum.

Example "break beam" counter code for Robolab and Mindstorms 2.0 can be found on our companion website – http://www.physicswithrobotics.com.

Note:

It should be noted that this is not the only, nor is it the best way to measure the performance of a ten second timer. Encourage your students to brainstorm several systems for measuring the timer's performance. The simpler the better! The process of designing and building this measurement system gives students excellent insights into how measurement tools work and specifically the issue of accuracy in a measurement tool. In the modern age, more and more measurement tools are "black boxes". We do not know how they complete their measurements yet we often trust the measurements they provide. In this section of the lab, students discover how challenging it is to design, build, and calibrate a measurement system.

Programming Ideas:

A system to measure the performance of a ten second timer can become quite complex. We encourage you to download and try out the code examples we provide on our website. For starters, however, we encourage you and your students to explore the most straightforward method. Figure 7.3.4 shows the Mindstorms code for this method. The NXT simply becomes an egg timer, beeping at the end of ten seconds.

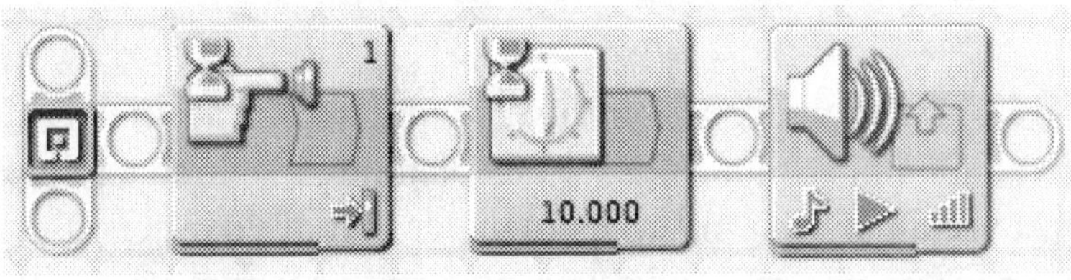

Figure 7.3.4 Mindstorms timer code. The timer starts when the button is pressed. After ten seconds, it beeps.

While this code does not represent the most robust solution to the problem of measuring a ten second timer's performance, it lays out the basic components of a measurement strategy. To measure the performance of the ten second timer, the system needs to know when to begin timing (the button in this case). It then needs to provide feedback to the user on the amount of time that has transpired. In this case, it beeps after ten seconds. To use this simple system, a student starts the pendulum at exactly the same time they press the button to start the NXT code. While the NXT waits for ten seconds, the student watches her pendulum. If she predicts that it will time ten seconds after five swings then she will watch for the pendulum to complete the fifth swing. If the NXT beeps as the pendulum completes the fifth swing then she knows her timer worked.

The simplicity of this system has a downside. It requires the student to manually start the timer and observe the end state of the pendulum when the beeper goes off. More advanced solutions are judged to be robust if they can take the human out of the equation all together. The time required to design this system is likely to be more than what is allotted for this activity in your physics curriculum.

Activity Instructions:

As you complete each of the following steps, document your work and record your observations. Refer to the table in the Student Observation and Analysis section of this activity write-up as a guide.

Part I

1. Using a collection of LEGO Beams, pegs, and LEGO weights, construct a physical pendulum.
2. Use the equation for the period of a simple pendulum to predict the period of your physical pendulum.

3. Compare your prediction with the results of your initial tests.
4. If necessary, change the equation for the period of a simple pendulum to better predict the period of a LEGO physical pendulum.
5. Create a new LEGO physical pendulum and predict its period. Test and compare with your prediction.
6. Return all materials to your instructor.

Part II

7. Develop a plan for creating a physical pendulum that accurately times 10 seconds with a whole number of swings. Your plan must include a request for materials. You must use all of your materials in the design of your pendulum.
8. Obtain the materials outlined in your plan and construct your "10 second timer".
9. Once your "10 second timer" is operational you are ready for Part III of this activity.

Part III

10. Create a RCX or NXT based system to test the accuracy of your "10 second timer". To ensure that your test is valid, your system must eliminate or minimize human error. See the notes and extensions section of this activity for programming tips.

Student Observations and Analysis:

As you proceed through the steps of this activity, record your work in the following table:

After you have constructed your physical pendulum, draw what it looks like in the box to the right.			
Use the equation for the period of a simple pendulum to calculate the prediction for the period of your physical pendulum. Show your work here.			
Test the period of your physical pendulum using at least three trials.	Trial #1	Trial #2	Trial #3
Describe how you will adjust (if at all) the equation for the period of a pendulum to better predict the period of a new physical pendulum.			
Create a new physical pendulum and draw it in the box to the right.			
Test the period of your physical pendulum using at least three trials.	Trial #1	Trial #2	Trial #3
Create a plan and a list of materials needed to make a "10 second timer" from LEGO Beams, pegs, and weights. Your timer must time 10 seconds in a whole number of swings. Document your ideas and materials list in the box to the right.			
Once you have created your "10 second timer", develop a plan to test it. Document your ideas for using sensors, a RCX or NXT, and a program to test the accuracy of your "10 second timer".			
Use the space to the right to document the results of your test. Conduct three trials.	Trial #1	Trial #2	Trial #3

Reflection Questions:

1. Describe your strategy for solving this problem. What did you measure/calculate in order to design an accurate "10 second timer"?
2. Summarize the relationship between the three variables (length, mass, and pullback distance) on the period of your pendulum.
3. When conducting an investigation of the relationship between variables, why is it important to only change one variable at a time?
4. In this activity, you did a short scientific investigation and then used the knowledge gained to design a practical device – a "10 second timer". What are other examples of instances where science helps engineers to design practical devices for society?

Notes and Extensions:

1. Students could do research on the difference between a simple and a physical pendulum to compare their results.
2. Although the period doesn't change the dampening effect of the pendulum, this can make readings more difficult. Have students find ways of making the pendulum more efficient in its swings.

7.4 Time for Robot Golf

Age Group: 14+

Essential Questions: Can you program a robot to score a hole in one?

Recommended Time: 45 to 90 minutes.

Materials:

NXT or RCX Kit
Measuring tape
Stopwatch
Blue painters tape[3] to make golf holes on the floor

Figure 7.4.1
Photo of Classroom Setup

Activity Overview:

In this activity, students will either build on previous experiences, or start from scratch to solve the problem. Students are going to play a round of robot golf. To solve the problem, they need to do the necessary measurements and calculations to predict the time the robot needs to travel to make it in the hole.

For simplicity sake, this activity is designed as a "hole-in-one" activity. For tips on setting up a more complex golf game, see the extensions section of this activity.

Learning Objectives:

By the end of this activity, students will be able to:

1. Interpret data to create an equation based or graph based mathematical model of an object's motion.
2. Use a mathematical model to predict the motion of an object
3. Test a prediction with an experiment
4. Refine their models based on new experimental evidence

---

[3] Blue painters tape is a type of masking tape and is suggested because it peels off of the floor or other surfaces easily.

Building Ideas:

If your golf match involves straight shots only, we recommend that you build a single motor vehicle (see chapter 2). If your golf match requires a robot that turns then build a two motor vehicle (see chapter 2).

Classroom set-up:

The number and arrangement of holes will depend on the time you have available for the activity and the design of your teaching environment. When the authors use this activity with students, they scatter the holes in amongst the lab tables of their rooms and place longer holes in the hallway. Shorter holes are on the order of 1-2 meters. Medium holes are 2-4 meters long and longer holes are 4-8 meters long.

Activity Instructions:

1. Build or use a previously built LEGO vehicle
2. Use Mindstorms or Robolab software to program your robot to travel at a certain power level for a certain time. You are now ready to start.
3. Consider the challenge posed to you by your instructor – "Figure out how much time to program your robot to play each hole in a game of golf". *Before* you start to play your round of golf, use a data table, graph, and/or equation to create a model of your LEGO vehicle's motion.
4. If you are going to vary the "power level" of your robot during this activity, be careful and think about the best way to collect and organize your data.
5. When you have finished collecting data on your robot, find the "practice green", which is 2-3 holes where you can practice before starting the game.
6. Measure the distance between the tee and the hole on the practice holes.
7. Use your mathematical model (data table, graph, and/or equation) to help you decide the time and power level to program into your robot so that it drives the distance to the hole.
8. If your robot misses the hole, measure the distance from its current spot to the hole and take an additional stroke to get it in the hole.
9. Based on your performance on the practice green, refine your model and start your game of golf.
10. Record your measurements on the provided scorekeepers card. (Table 7.4.1)

Hole	Distance to Hole	Robot Power Level	Time Calculated	Strokes
1				
2				
3				
4				
5				
6				
7				
8				
9				
			Total Strokes	

Table 7.4.1 Scorecard

Student Observations and Analysis:

1. Use a data table (Table 7.4.2) and graph to create a mathematical model that allows you to accurately predict how to operate your vehicle to get the best score possible!

Table 7.4.2 Robot distance and Time Data

Time of Travel	Distance Travelled	Notes

2. From the data above draw an appropriate graph to determine any distance from the time.

Reflection Questions:

1. Describe how you played each hole. What steps did you follow?
2. Did you use the same process for each hole? Were any holes better than others? Explain.
3. Were there any errors in your process that you had little control over? Did you create a strategy for dealing with these errors?
4. To complete this game of golf, you applied your knowledge and skills in science, math, and engineering. Is this true for all athletic games? Pick an example and explain.

Notes and Extensions:

1. If you are using the RCX or the NXT with Robolab, pay attention to how well the vehicle goes straight. Encourage students to consider this as a critical design feature. Single motor cars and cars with four wheels and a wide wheelbase are good suggestions to students. Two motorcars that do not drive straight are a source of great frustration for students when they want to perform well in this activity.
2. Depending upon the abilities of your students, you may be interested in increasing the challenge of the activity. A few ideas for increasing the challenge are:
   a. Changing the size of the holes. Extra points can be awarded for a hole in one on smaller holes.
   b. Add "sand traps", "dog legs" or other obstacles to move around. This offers the opportunity for students to practice vector analysis skills. For a hole with obstacles, students are allowed to measure the straight-line distance between the tee and the hole. Then, they must predict two or more displacement vectors that will put the robot in the hole.
   c. As an extra programming challenge, students need to create a program to move the robot along their predicted displacement vectors.
   d. Speed round. Students must complete as many holes as possible in a given amount of time.

7.5 Design a Motorised Crane Challenge

Age group: 14+

Recommended Time: 2 hours

Challenge: Can a light LEGO based crane be made strong enough to lift more than its own weight?

Materials:

LEGO pieces	RCX brick	LEGO Motor
LEGO gears and pulleys	string various	brass weights 10 - 100g

Activity Overview:

This is a follow up engineering challenge to the gear ratio investigations.
It is used to apply their knowledge of gear ratios and force along with construction skills. Some understanding of ties, bracing and rigid structures is also required.

Learning Objectives:

By the end of this activity students will be able to apply the following concepts:

1. Simple machines use pulleys and gears to reduce force.
2. Gantry structures consist of rigid frames, braces, struts, ties and cantilevers to balance weight.
3. Efficient energy conversions require reduction of friction and movement.

Engineering Challenge

Plan, design and build a motorised LEGO crane to lift as much weight as possible.

Building Instructions and Conditions:

1. You are limited to no more than 450g of LEGO, 1 motor, 1 RCX/NXT Brick and 1 metre of string.
2. The crane must be a minimum of 20cm in height (base to the arm)
3. This crane arm must extend a minimum of 20cm (base to string) to raise weights at least 50cm using an electric LEGO motor. (See Figure 7.5.1)
4. The base can be fixed to the desk or held down.
5. The motor has to be controlled by a 9V battery source. The RCX or NXT Brick can supply this. Battery source and motor can be fitted anywhere on the structure.
6. The weight of LEGO used is limited to 450g (not inc. RCX and motor)
7. You can use devices to aid lifting including pulleys, gears and string.
8. String can also be used to stabilise the structure. You are limited to 1 m of string altogether.

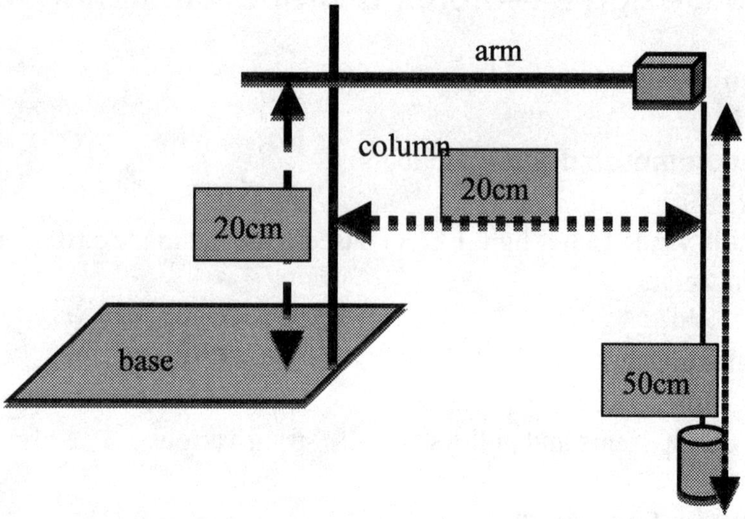

Figure 7.5.1 Diagram of specifications

Activity Instructions and Assessment:

There are three parts to this assessment. You need to submit:

1. The Design and Construction of the Crane must include a folio, which shows a progression of draft diagrams. (See Table 7.5.1) Diagrams should be labelled showing Forces to be balanced.
2. Results of investigations of different scientific tests you have done to increase weight lifted.
3. Report on your final outcome, minimum 250 words. This must include:
   a. Your Best Weight Score = Weight Lifted ÷ Weight of Crane
   b. Explanation of how you kept the Crane Stable.
   c. Explanation of how you reduced energy losses to make the lifting more efficient.
   d. Overall how effective was your Crane and what limitations would need to be overcome for a full-scale application?
   e. Attach your folio notes.

Notes and Extension:

1. The conditions can be changed but it is important to limit requirements, so that students need to improvise and use physics knowledge such as the function of pulleys or levers to reduce effort.
2. It is very important for students to design and test first. This includes:
   - Drawings submitted.
   - Different Lego structures tested for strength.
   - Motor and pulley tests.
   - Changes recorded and justified.
3. Factors such as motor power levels, speed of elevation, thickness of string etc. could also be tested.

My Name: _____________Team Members Initials: __________________

Design Ideas and Testing

Use sketches and descriptions to document your new ideas.	**Explain why you believe your new ideas will work better.**	**Summarize the results of evaluation measurements (i.e. what weight it lifted, where it failed)**	**Describe any changes you will make to your next design**
Design #1			
Design #2			
Design #3 *If you need additional space to document the results of tests on additional designs, use extra paper and attach it to your lab write-up.*			
Final Design			

Table 7.5.1 Sample Folio outline

Criteria for Marking Sheet

Elements:

Knowledge: Forces identified, Energy losses identified

Investigation: Plan Investigations. Design. Conduct tests. Identify problems. Draw conclusions

Communication: Explaining function. Clear diagrams

Reflection: reflect on learning and adjustments for better outcomes.

Criteria	**A**	**B**	**C**	**D**
Knowledge:	Comprehensive Knowledge and appropriate application of the task to science concepts learnt.	Thorough Knowledge and application of the task and Physics involved.	Satisfactory Knowledge and application of the task and Physics involved.	Limited Knowledge and application of the task and Physics involved.
Design, Test and Modify	Detailed evidence of progression in designs.	Evidence of some progression in designs.	Designs presented	Some design effort
Construction:	Results show thorough and fair testing	Results show fair testing	Some Testing done	
	Applied test results well to complete an effective and very efficient product.	Completed an effective and efficient product.	Completed a product that is efficient.	Completed a product that works.
	Lifting Score > 2	Lifting Score > 1	Lifting Score > 0.5	Lifting Score < 0.5
Communication: In explanation of how construction is stabilised and made to increase efficiency, the student	Explains accurately and clearly how their design works, using correct scientific language and detailed diagrams. Modifications clearly explained.	Explains clearly their design and uses correct scientific language and clear diagrams to illustrate. Modifications explained.	Explains their design. Uses diagrams and scientific language. Evidence of Modifications	Explanation and diagrams attempted.
Reflection	Insightful answers to effectiveness of product and adjustments required for a more effective working crane.	Correct answers to effectiveness and limitations of crane.	Answers show some reflection of limitations	Limited or no reflection.

Chapter 8

Projects

8.1 Automated Control Project – An Engineering Challenge

8.2 Study of Real Motion – An Open Investigation

8.1 Automated Control Project

Age group: 15+
Time: 3-4 weeks
Programming Skills: Advanced MINDSTORMS or ROBOLAB.

Introduction:

This project has been given for the last few years in a senior Engineering class and forms the culminating assessment piece for a contextual unit in year 12 Physics on control systems. Students will have studied electronics and electronic systems, have some experience in programming (MINDSTORMS and PICAXE Basic) and have done some theory and practice on closed systems using simple feedback and proportional control. This control systems unit is done in one term (10 weeks) and this project would be given towards the end of the unit.

Task Description:

You are required to produce a working model or device using control technology. The model or device must have a real application. It must use at least one input and one output device, be managed by a programmable microcontroller and involve some consideration of materials and processes.

There also must be some element of design. This can be a circuit with appropriate peripheral devices or a dynamic microcontroller such as the LEGO NXT/RCX incorporated into a closed control system.

Examples:

Computer controlled hoist.
Conveyor system.
Traffic Control system
Model smart car park
Safe amusement ride
Energy Efficient House
Working Solar System model
Sentry System
Gate Timing system for measuring high-speed model racers.
Quiz Show Buzzer System
Aquarium temperature/ pH controller
Robotic Arm for a particular purpose.
Sensing Device that monitors coming and going in a room.
Smart Bridge system. For example: a sensor based Drawbridge.

Project Assessment Criteria:

- Appropriateness of your module to specified goal and societal needs.
- Complexity and ownership (how much you designed yourself)
- Understanding and knowledge of component systems and circuit design.
- Understanding of programming and software requirements
- Investigating effectiveness and explaining choice of all major components (materials, circuit and components, input and output devices, mechanics etc.)
- Construction of the electronic circuits and /or LEGO modules.
- Written justification of the design and the way it was improved to meet specified goals.
- Communication of the above items by appropriate diagrams, programs and brief explanations. (This includes Diary Log Entries).
- Performance of device or model.

See Page 186 for a sample rubric.

Length:

Must include at least 1000 words of written analysis plus diagrams and specifications for building materials.

Criteria and Standards Schema for Automated Control Project

Criteria		Standards				
Objective	Outcome	A	B	C	D	E
Knowledge						
Collecting, organising and using data for a successful construction.	Knowledge and skills in construction and assembly.	Excellent demonstration of working knowledge and skills in the process of the design and construction.	Good demonstration of working knowledge and skills in the process of the design and construction	Satisfactory demonstration of working knowledge and skills in the process of design and construction	Some demonstration of working knowledge and skills in the process of design and construction	Little demonstration of working knowledge and skills in the process of design and construction
	Details knowledge in explaining workings of mechanical and electrical systems.	Includes a large number of core engineering and programming concepts.	Includes a fair number of core engineering concepts.	Includes a few core engineering concepts.	Includes some core engineering concepts	Fails to demonstrate much knowledge.
Reasoning						
Elements of Design	Proposes an outcome and design that fulfils specifications.	Design meets specified outcomes. System requirements of inputs and outputs are met.	Design meets most specified outcomes. System requirements of inputs and outputs are met.	Design meets some outcomes. System requirements of inputs and outputs are attempted.	Design attempted but few requirements met	Little or no elements of design shown.
Investigation and Analysis.	Tests components and determines most effective usage	Correctly measures and test all aspects. Tests and adjusts components for most effective usage.	Measures and tests most aspects. Tests and adjusts components for effective usage.	Does some measuring and testing of components.	Little time spent testing project.	No testing time done.
	Redesigns and modifies to improve performance	Demonstrates good reasoning and adjustments to improve performance with success.	Reasoning and adjustments evident to improve performance with some success.	Some attempts evident to improve performance	Little evidence to improve performance	No evidence.
Justification	Justifies reasons for component choices.	Clearly justifies all aspects of design in terms of performance and safeguarding against possible problems.	Justifies most aspects of design in terms of performance and safeguarding against possible problems.	Justifies some aspects of design in terms of performance and safeguarding against possible problems.	Justifies few aspects of design in terms of performance and safeguarding against possible problems.	Does not justify reasons for choice.
Communication						
Communicate information in various contexts	Presentation of folio	Diagrams and pictures are clear and in appropriate logical order. Overall presentation is of a very high standard.	Diagrams and pictures are generally clear and in appropriate logical order. Overall presentation is of a high standard.	Diagrams and pictures are present and easy to identify. Overall presentation is of a satisfactory standard.	Some diagrams and pictures present. Overall presentation is a low standard.	Little attempt has been made to display diagrams and pictures. Overall presentation is poor.
	• Text	Text is well presented with clear explanations and language.	Text is well presented with mostly clear explanations and language.	Adequate explanations and language	Careless errors and poor explanations	Little explanation made.

8.2 Open Investigation into Systems of Real Motion.

Age group: 15+
Time: 4 weeks/ 8 lessons

Introduction

In Senior Physics in Queensland, we have a contextual unit on real motion. The students study amusement park rides and drag cars. By the end of this unit they have an extensive understanding of Motion, Forces and Energy (about 12 weeks). As a culminating assessment, they are to do an "Extended Experimental Investigation" or EEI.

This includes devoted class time as well as written work at home. Students have chosen a variety of investigations, usually working together in pairs for the practical work. They may choose data logging or video analysis or both. They have done many of the activities given in this book as part of their study work, so are familiar with the skills and equipment.

I am amazed at the depth of study. A student will explore when they are motivated by an investigation of their choice. One group studying the swing versus flight of a golf ball extended their study into the path of a dimpled golf ball, looking at the causes of the extra distance. This depth can never be gained in a standard class lesson.

Many of the example projects shown on page 188 could be modelled using the NXT or RCX. The activities of this book will also give students good "starting points" as they come up with their own ideas for an investigation.

Your Task: Investigating Real Motion

- Choose an interesting mechanical system where motion and energy can be predicted and measured.
- Investigate the motion and energy changes and the effects of force on this motion.
- Hypothesise the factors affecting the outcomes of the motion (at least two) and verify/refute this in experiments.
- Experiments should involve data logging, by either sensor analysis or video analysis.
- Analyse your results using software such as Logger*Pro*, MINDSTORMS 2.0 or ROBOLAB.
- Use the results of your experiments to conclude optimum conditions to maximize outcomes of this system.

Conditions and Requirements

- Determine your Investigation and what your equipment requirements are by the end of week 1.
- Complete all trials of your experiments by the end of week 3.
- You must keep a daily journal of all your notes and results.
- Submit a draft of your results and data analysis by the beginning of week 4.
- Submit a draft of Introduction and Conclusions by the end of week 4
- Submit your final report by the Due date, which must include journal, introduction, hypothesis, experiment outlines, data collection, data presentation, data analysis including calculations and justification of various conclusions made. (See Criteria sheets for how these are marked)

Examples of possible mechanical systems to investigate:

Model Roller Coaster motion
Flight of model rockets
Energy loss in a vibrating spring
Golf swing and ball motion.
Investigating jumping or running and impact on the feet
Motion of the centre of gravity of irregular objects
Measuring resistances in Bicycle motion.
Resonance of a spring/mass system
Factors affecting flights of a projectile
Factors affecting performance of a model drag car
Factors affecting energy loss for a bouncing ball
Efficiency of an electric motor
Efficiency of pulley systems
Energy losses in Collisions
Trampoline Motion
Mechanics of a tennis serve
Forces on a bridge as a vehicle traverses

CRITERIA ASSESSING **Real Motion Investigation**

Gen. Obj.	Element	A	B	C	D	E
Knowledge and Conceptual Understanding	**1. Reproduction and interpretation of concepts, theory and principles (Introduction)** **2. Analysis and calculations.** **3. Discussion - comparison and explanation of concepts**	**1. Introduction** is concise, engaging & accurately describes project **2. Analysis** clearly links and accurately applies correct formula to explain challenging and complex concepts & phenomena **3. Discussion** shows excellent understanding of challenging complex concepts & principles	**1. Introduction** is succinct, interesting & adequately describes project **2. Analysis** clearly links and uses correct formula to explain most concepts & phenomena **3, Discussion** shows understanding of some challenging or complex concepts & principles	**1. Introduction** is brief & informative **2. Analysis** clearly links and uses correct formula for straightforward situations and simple concepts. **3. Discussion** shows understanding of concepts & principles.	**1. Introduction** is brief or informative **2. Analysis** applies some simple algorithms for simple concepts. **3. Discussion** reproduces simple concepts & principles	Lacks Understand-ing throughout
Investigative Processes	**1.Management of research task (Literature review & Bibliography)** **2.Design of Practical Investigation Intro, Planning, Aim Hypothesis** **3. Recording of Risk, Equipment and Data, (Method, Results)** **4. Analysis and Discussion of data - Explanation of Errors/Anomalies**	1. Research has been thorough and expressed in an appropriate format and to an appropriate length. The **Bibliography** shows valid and wide sources, correctly referenced. 2. The **Intro, Planning, Hypothesis & Aim** generate and justify a significant question with clearly testable hypotheses, leading to a detailed, efficient, well executed experimental design. 3. The **Method** and **Equipment** are successfully managed to safely produce valid **Data** with credible observations. 4. The **Analysis & Discussion** show accurate and systematic analysis to identify relationships between patterns & trends, errors and anomalies	1. Research has been reasonably extensive and expressed in an appropriate format. The **Bibliography** shows valid or wide sources, correctly referenced. 2. The **Intro, Planning, Hypothesis & Aim** generate and justify a question with a testable hypothesis, leading to a well executed experimental design. 3. The **Method** and **Equipment** are successfully managed to safely produce valid **Data.** 4. The **Analysis & Discussion** show accurate analysis to identify patterns and trends, errors and anomalies	1. There is evidence of research and it has been expressed in an appropriate format. The **Bibliography** is suitable with some, correct referencing. 2. The **Intro, Planning, Hypothesis & Aim** generate a question and hypothesis, leading to a satisfactory experimental design. 3. The **Method** and **Equipment** are managed to safely produce appropriate **Data.** 4. The **Analysis & Discussion** show analysis to identify obvious patterns and trends, errors and anomalies	1. There is little evidence of research and/or this is reported in an appropriate manner. The **Bibliography** is present 2. The **Intro, Planning, Hypothesis & Aim** describe a given question leading to an experimental design 3. The **Risk** and **Equipment** are managed to safely produce **Data.** 4. The **Analysis & Discussion** identify obvious patterns and anomalies	Poorly conducted and analysed

Gen. Obj.	Element	A	B	C	D	E
Evaluating and Concluding	**1. Conclude & evaluate the hypothesis based on results.**	**1.** The **conclusion** clearly answers the hypothesis. Expected and observed results are appropriately compared and evaluated.	**1.** The **conclusion** generally answers the hypothesis. Expected and observed results are compared and evaluated	**1.** The **conclusion** responds to the hypothesis. The observed result is clearly presented.	**1.** The conclusion does not clearly respond to the hypothesis. The observed results are not clearly presented.	Little evaluation evident
	Refinements evident.	Realistic refinements proposed.	Refinements referred to.	Some evaluation and refinements evident.		
	2. Select & Present data (Diagrams, Tables & Graphs -	2. The **Presentation** shows discriminating selection and display of data using a range of innovative formats.	2. The **Presentation** shows suitable selection and display of data using a range of formats.	2. The **Presentation** shows suitable selection of data and appropriate display.	2.The **Presentation** shows some suitable data.	Poor Presentation and Communication evident.
	3. Overall Communication	3. The **Overall communication** is fluent, cohesive & grammatically correct.	3. The **Overall communication** is cohesive & grammatically correct	3. The **Overall communication** is cohesive.	**3. Overall communication** shows some cohesion	
	4. Detailed Journal Showing progression.	**4. Journal** shows a regular, detailed log of activities and changes.	**4. Journal** shows a regular, log of activities and changes.	**4.Journal** shows a log of activities.	**4. Journal** shows a log of some work done.	